# THE RADICAL MIDDLE

## HOW FOSSIL FUELS AND RENEWABLES CAN COLLABORATE ON THE FUTURE OF ENERGY

DAN ROMITO

Paperback ISBN: 979-8-218-62472-9

Printed in the United States of America

First Printing, 2025

*Thank you to all those who convinced and inspired me to keep moving forward – you know who you are.*

# CONTENTS


1. Green Ain't Clean 1
2. The Vilification of Energy Companies 19
3. Us vs Them Doesn't Work 35
4. The Radical Middle Mindset 51
5. Meeting in the Radical Middle 67
6. How to Be Curious about Energy Stakeholders 79
7. How to Be Curious About the Environmentalists 93
8. How to Be Curious About Policymakers 107
9. Staying the Course When Opposition Mounts 121
10. An "All Of The Above" Set Of Energy Solutions Will Make the World a Better Place 133

Notes 143

About the Author 149

# CHAPTER 1
# GREEN AIN'T CLEAN

Energy is the lifeblood promoting a respectable standard of living. Affordable and reliable access to power paves the way for necessities and luxuries modern societies often take for granted. This is especially true in Western economies. From hot water to refrigeration, transportation to toasters, Wi-Fi connectivity, and much more, everyday conveniences are all powered by reliable energy. Deprive any society of energy, and quality of life plummets. For the vast majority of societies around the world, quality of life means using fossil fuels.

The energy industry has become a hotbed of ideological debate, which doesn't help the energy transition. People are caught in the misperception that to hasten an energy transition, the energy sector needs to switch over to renewables and cut out fossil fuels completely. But in truth the misunderstandings—about how energy works, renewable capacity, and all-or-nothing thinking and policy—pit stakeholders against each other, and make solutions to our energy problems all the more difficult to solve. Energy companies, environmentalists, and policymakers are so entrenched in their ideas of how energy ought to work (rather than how it actually works), they struggle to make marked progress in solutions for the myriad problems the energy industry faces in the decades to come. These groups need a better understanding of how energy works and how to communicate effectively to utilize all the resources and solutions available.

On a 110°F day in September 2022, Elliot Mainzer, CEO of the California Independent System Operator (CAISO) experienced this dichotomy. A record-shattering heatwave pushed the state's electric grid to the brink. With temperatures soaring to dangerous levels, air conditioners across the state were sucking up power at an unprecedented rate.[1]

Elliot knew there was only one possible solution to keep the lights on: ramp up every available power source. It was the only way to bridge the gap between supply and demand. But this meant leaning heavily on fossil fuels—an inconvenient truth contradicting California's ambitious but unrealistic decarbonization goals.

It was a paradoxical predicament. To avert an energy crisis, Mainzer had to lean on the resources California was trying to phase out in pursuit of ambitious renewable goals. Wind and solar power supplied over a third of the state's electricity, but that wasn't enough when energy demands soared. Mainzer had no choice. Without on-demand power from natural gas turbines to balance out supply when renewable generation dipped, the whole system would collapse, jeopardizing the health and safety of millions of residents.

At the core of this standoff was a dangerous misperception: the notion that rapidly eliminating fossil fuels was the only path to decarbonization. In reality, every state in the United States will need fossil fuels (largely natural gas) to facilitate the transition to renewables while still keeping the power on. In Elliot's situation, he knew a more nuanced approach was required.

As the day unfolded, anxiety skyrocketed across Elliot's teams. Employees worked tirelessly at CAISO's headquarters, power plants, and grid control centers. Had they accurately estimated power demand? Would aging transmission lines hold up under the strain? One misstep could trigger a cascade of blackouts.

Finally, to stave off such a catastrophic failure, Elliot fired up idled natural gas plants and issued statewide emergency alerts urging consumers to cut their power usage. Governor Gavin Newsom jumped in, too, texting Californians to turn up their thermostats.

Utility workers, grid operators, power plant technicians, and emergency responders stood ready to act if equipment failed or power lines went down. The whole system was stretched impossibly thin.

But as evening fell, an unlikely event occurred. Californians answered the call for conservation and cut their power use by over 2 gigawatts, the rough equivalent of turning off 200 million LED lightbulbs. This made all the difference. CAISO squeaked by without major outages thanks to the combination of the above-mentioned power generation strategy and individual residents' efforts.

Near misses like this have increased over the last decade, highlighting the role of fossil fuels as an essential backstop for energy grids nationwide. But this issue only reflects the broader challenge facing the energy sector, policymakers, and consumers: How do we incorporate "clean" power without sacrificing reliability and raising costs. In seeking to eliminate fossil fuels at a break-neck pace, California created a grid reliant on technology unable to cope with surges in energy use.

Mainzer's white-knuckle experience shows the risks of treating fossil fuels and decarbonization as an either-or proposition. The two are profoundly and perpetually interlinked—for the foreseeable future, we need oil and gas to enable the shift to cleaner energy by providing always-on power to fill in the gaps. Cleaner energy and fossil fuel are not mutually exclusive. Axing fossil fuels without viable and reliable alternatives in place is a recipe for blackouts, backlash, and regression in standard of living.

All economies, particularly developing economies, pursue energy in a cheap, reliable form. Yet, as the world develops with rapidly evolving technology, energy demands continue to rise. And because the most gains will go to the most developed economies, developing nations are in an energy arms race. Possible negative externalities are of little consequence in the pursuit of growth.

But the "negative externalities" are not only possible, they are—and always have been—a reality. Burning fossil fuels for energy produces carbon emissions. That much is undeniable. Because of increasing concerns over emissions' effects on the globe, there are widespread efforts by developed countries to pursue cleaner forms of energy and reduce emissions.

Historically, the expansion of economic output (usually measured in Gross Domestic Product per capita, or GDP) requires the expansion of an emissions profile. However, over the last twenty-five years western economies have figured out how to reduce emissions without sacrificing economic prowess. This progress has led many members of these developed countries to believe all nations should join in these efforts to transition as quickly as possible to renewable or "green" tech.

It is these calls to action many policy makers and leaders of developing nations take issue with. In most developing nations, green tech is unaffordable and cannot generate the amount of power the country needs to maintain GDP per capita, let alone expand it. To prioritize decarbonization would mean a significant drop in the standard of living for these countries. And any policy sacrificing standard of living for decarbonization is not sustainable.

Change in the energy sector is tricky because it requires energy companies, policymakers, municipalities, and environmentalists to be on the same page. Right now, they're not even reading the same book. Each group inhabits and consistently endorses its own echo chamber, intentionally oblivious to how the others' industries actually work. Too often, debating parties pass judgment and push unrealistic solutions that sound good but don't work in the real world.

The most significant disagreement in the energy policy space is the push to eliminate fossil fuels. People think fossil fuels counter energy innovation and decarbonization efforts. This is false. Fossil fuels have always been a driver of renewable technology and an efficient means of executing decarbonization strategies. We've been conditioned to believe we must abandon oil, coal, and gas to "go green." But, the reality is, we need fossil fuels to maintain living standards and meet the increasing energy demand, not just abroad, but in the United States.

But what if there was a way to decarbonize the global economy without sacrificing geopolitical standing or tanking standards of living? It is possible, but only by strategically optimizing and accepting the long-term utility of fossil fuels. Unfortunately, this means asking sworn ideological opponents to align on their extreme differences of opinion.

The energy sector needs a concerted effort, a collaborative approach that brings diverse groups together to adapt to soaring demand and environmental concerns. The potential of this collaboration is immense, and it's something nearly everyone I speak to—from engineers to executives to representatives to lawyers and activists—can see. They all lament how difficult it is to work with the people in the other camps, but they also know the potential for change. Some however, are less aware of how the whole energy system works, which can be a barrier in the most basic conversations…

## ELECTRICITY DOESN'T COME FROM ELECTRICITY

The energy grid is complex and often misunderstood. Many people might see increasing numbers of electric devices—from induction stoves to electric vehicles (EVs)—and think we can remove fossil fuels now. There's so much innovation happening it's easy to assume change in the energy sector should happen quickly. However, policymakers, environmentalists, and even energy sector professionals often overlook *how* we generate the electricity running our homes, offices, and public facilities. It's easy to call for clean energy when we don't understand the complex nature of energy and how reliant we are on fossil-fuel-generated electricity.

While traveling the world to discuss the energy sector, I've discovered most people don't know where the energy powering their electric devices comes from. People think of electricity in an isolated manner, and while some electricity can be "clean," to know if it is clean or not, we need to understand the fuel providing that electric energy. For example, an induction stovetop uses electricity, but it could come from a power plant burning coal.

It's common to forget electricity comes from somewhere else. Even some of the most intelligent, successful individuals among us are unaware.

A few years ago, I gave a talk to a high-powered group of executives, energy policymakers, and lawyers—you know, a crowd known for their humor and joviality. As I wrapped up my presentation, I could hear people rearranging themselves in their seats. It's typical for me to receive questions from upset individuals after my talks, even though it is simply a collection of charts, graphs, and tables.

A woman in a sleek, tailored suit eyed me skeptically as I wrapped up. Within seconds of opening the floor to the audience, she held a microphone. Preparing to speak, she removed her glasses and perched them on top of her white-gray hair.

"Hi," she began, "my name is Barbara, and I'm from D.C.. I was just curious as to why we can't electrify everything. It seems like we're so close, we just need to eliminate fossil fuels to make room for more environmentally friendly initiatives."

"Sure," I nodded calmly, knowing exactly where good ole' Barb would take the conversation. "I'm glad you brought that up, Barbara. Electricity is great, but the electricity must come from somewhere." I paused and let my statement sink in, then asked: "Where do you think our electricity comes from?"

Her eyebrows arched, taken aback by the question. "What do you mean? Electricity comes from...electricity."

This is where I typically feel the start of a dull pain above my left brow, begging me to crawl into a deep and dark hole because my faith in humanity is vanishing rapidly.

"Well, let's say your home is all electric—what is creating that electricity?"

"Electricity," she said. She answered it the same automatic way people say "blue" when asked about the color of the sky.

There were others queued up behind her, so I didn't have time to guide her to the correct answer—I had to spell it out for her.

"Barbara," I started as gently as I could, "electricity does not come from electricity. It must be generated." I turned to the crowd. "Does anyone want to guess how most of the power in the States is generated?"

No one jumped in. Barbara blinked at me, her confidence wavering.

"In the U.S.," I began, addressing the whole room, "41 percent of our power is generated from natural gas plants. Coal accounts for nearly 16 percent Solar-generated power is about 9 percent and wind is around 11 percent." I turned my attention back to Barbara. "It's a tough pill to swallow, considering our environmental ambitions."

Barbara raised an eyebrow. "Well, I don't know if I can trust your data," she said into the microphone.

I held in a scream of frustration.

"Okay, Barbara." I turned my attention to the crowd again. "But just so everyone's clear, all the data I shared today are from the US government's Energy Information Administration. It's all publicly available. Anyone can look at it, though most people don't."

Barbara curtly thanked me for my time and sat back down. But I could tell she wasn't sold.

I answered a few more questions from the crowd until time was up. I wasn't able to catch Barbara afterwards—she was nowhere to be seen.

Later, back in my hotel room, I mulled over the interaction. If Barbara—an accomplished, intelligent, experienced professional—didn't know where electricity came from, how many more people thought the same? It is a question and topic that comes up almost every time I speak to a group of generalists. To think there were people in every audience like Barbara, unaware of the basics of power generation—the dull pain behind my eyebrows was now a full-blown headache. To her credit, Barbara had been brave enough to admit it and ask questions, and though she didn't seem particularly swayed by my answers, at least she spoke up.

We may have more electric devices than ever—from laptops to induction stovetops to underfloor heating to EV charging ports, but our *power sources* have been slower to change. Electricity doesn't just come from electricity—it must be generated from something. It's a fundamental law of thermodynamics: energy cannot be created or destroyed, only converted from one form to another. Without getting too into the weeds on thermodynamics laws (we'll leave that to the engineers), a motion or chemical reaction produces heat, which can be converted into energy. It is literally impossible for electricity to produce more electricity).

Even in a developed nation like the U.S., most of our power comes from fossil fuels, though each state has a different power generation mix. Consider three different states: Washington, Illinois, and Virginia. Each has its own power needs and sources.

Washington, thick with forests and raging rivers, generates the majority of the state's power (around 67 percent)[2] through hydroelectric. Massive dams on the Columbia River Gorge and the state's extensive rivers and mountains provide ample opportunities for hydroelectricity. Of course, Washington fills in the gaps with natural gas, wind, and solar.

Illinois, by comparison, largely relies on nuclear power. In 2023, 54.9 percent of the state's power was produced by several nuclear-powered plants.[3,4] The rest of the state's power was generated from natural gas (around 16 percent), coal (around 15 percent), and renewables (nearly 14 percent, mostly wind).[5] Illinois is doing so well in energy production thanks to its large nuclear plants, the state sells excess power to neighboring states. (Nuclear power is often shunned as an energy source because of its potential dangers. However, it is reliable, creates massive amounts of power, and is cleaner than alternatives like coal. Nuclear is so reliable, it powers most of the U.S. Navy.)

Then there is Virginia. A state which historically ran off of coal, Virginia now only uses coal to generate about 4 percent of its power. Natural gas and nuclear energy dominate the generation, at 54 percent and 31 percent, respectively.[6] Biomass and solar, "green" alternatives, contributed 11 percent of power generation, and petroleum made up the difference.

However, while Virginia doesn't run off coal anymore, it continues to export it. In 2022, Virginian ports "handled 38% of the nation's coal exports."[7] Virginia doesn't use coal anymore, but it's still a valuable option for energy generation around the globe. (In 2023, coal accounted for about 35.5 percent of the world's electricity generation, making coal the largest source of electricity in the world, followed by natural gas, which produces about 22 percent of the world's electricity.)[8]

Electricity doesn't come from electricity and depending on the state, there is a varying mix of power sources. Each of these states has a different makeup of power generation because optimal energy mix is contingent on a state's resources and energy demands. Virginia has recently been seen as a location for companies to host massive data servers running AI and cloud computing. This will cause a massive demand on Virginia's current aging grid. The introduction of these energy-intensive data servers cause concerns about the cost and impact of the transition to cleaner energy sources, since renewables generate less power than traditional fuels.[9] If Virginia can't come up with a long-term, viable answer to higher demands on their power, residents of the state will face the consequences in the form of power shortages, higher costs, or both.

It may sound obvious that electricity must come from somewhere, but many people overlook where the power in their home comes from. Residents of an entire electric home or employees in an electrified office might think they are running on clean energy—but depending on where they're located, their electricity might be generated by fossil fuels. Getting rid of those power sources in favor of clean sources will not happen overnight, particularly as the rise of technology (e.g., data centers and chip factories) demands more power usage.

Change in the energy sector is slow because we've discounted fossil fuel-generated power's role in the industry and pitted innovation against fossil fuels. Many people thought we could give up fossil fuels when the first renewables entered the scene. But while America ramps up its renewable efforts, every state is discovering it still needs fossil fuels for a stable grid.

In America, it would require a herculean effort to halt the use of fossil fuels and rely solely on "clean" renewables. Energy consumption would need to drastically reduce energy consumption which would impact the US's economic output and heighten geopolitical tensions. Even if America substantially decreased energy consumption and switched over to all renewable power sources, there is still the energy hungry globe to consider.

## ENERGY HUNGRY NATIONS

As America innovates to lower emissions, global energy consumption increases, putting immense strain on power grids worldwide. India and China, the world's production and population hubs, require massive amounts of power to fuel their economies. As a result, their carbon footprints are drastically expanding.

India and China have insatiable appetites for energy. In 2023, China consumed 47,428 terawatt hours of energy, with India trailing behind around 10,838 terawatt hours. Combined the two nations consume double the amount of energy consumed by the U.S. (26,189 terawatt hours).[10] By 2035, China and India combined are projected to account for *half of global energy growth,*[11] consuming 31 percent of the world's energy, up from 21 percent in 2008.[12] As these nations develop, the billion citizens each has will demand a higher standard of living—of which reliable energy is crucial. India and China have no choice but to quickly and reliably provide power to agriculture, businesses, and the average household.

There is nothing wrong with expanding access to energy, however, the issue is China and India's primary energy sources are the dirtiest of the fossil fuels (i.e., coal and oil). This is because these are the cheapest sources, and China in particular has seen success mining coal domestically. Renewables are often too expensive for these countries and do not provide the steady baseload power required. In China, coal and oil still account for over 70 percent of total energy consumption, creating an immense amount of annual emissions. The United States may consume vast amounts of energy, yet the U.S. is responsible for less than 15 percent of global carbon dioxide emissions—India's and China's emissions combined comprise about 40 percent of global emissions.

While the U.S. is blessed with cleaner power sources like hydroelectric dams and the capital to invest in the latest renewable tech, many other countries don't have that luxury yet and, frankly, might never. Many countries are still in the process of building out their core energy infrastructure and power grids. They need steady, low-cost power *now* to industrialize and lift their populations out of poverty. Forcing an overnight shift to expensive, intermittent renewables is not feasible.

Energy debates between energy companies, environmentalists, and policymakers often center around how quickly the world can phase out fossil fuels. But given the projections of energy use in the coming years, fossil fuels will remain a significant part of the global energy mix for decades. It is an empirical, political, scientific, and economic certainty that they will remain and not fade anytime soon. Eliminating all fossil fuels is not an option, but we should use fossil fuels as cleanly and efficiently as possible to enable rising living standards while minimizing environmental damage.

Ironically, the ideology skewing the energy debate is actually delaying the energy progress of developing economies. When affluent countries try to force the developing world to use only green energy, this can slow the developing nations' ability to modernize, directly impacting quality of life. The hard truth is global energy policy and innovation efforts face powerful geopolitical headwinds, and one-size-fits-all policies and solutions fail to factor this in.

## ENERGY POLICY DOESN'T EXIST IN A VACUUM

The U.S. energy sector is part of global economics and is slow to change because one shift in the industry affects the global economy. The entire world runs off of fossil fuels. Fossil fuels power economies and provide people with basic standards of living. The U.S. energy sector has been slow to weed out fossil fuels because the nation needs them to compete on the global market and as a geopolitical tool. Releasing more oil onto the global market can decrease the global price of oil, preventing oil giants like OPEC from having too much ecopolitical power.

When it comes to innovating the energy sector, the U.S. cannot ignore the complex geopolitical and economic forces at play globally. Pushing simplistic fossil fuel bans without accounting for development needs is intellectually dishonest and will only breed resentment and non-compliance. We need a more sophisticated, collaborative approach to cleaning up the world's energy systems that balances economic and environmental priorities. There are no easy answers, but burying our heads in the sand is not an option. Fossil fuels are here to stay for now, and so we must figure out how to leverage them as a tool for our own energy needs, the globe's, and as a means to stabilizing political forces.

The energy sector is additionally slow to change because domestic energy policy has global effects. Stop coal production in Virginia because of clean energy protocols, and Virginia's environmental outcomes might improve, but there will be ripple effects on the U.S. economy and the global environment. Because the U.S. is a major powerhouse economically and politically (and in energy), the federal government's actions, even a state's actions, ripple out to affect citizens in other countries.

To carry on the coal example, if U.S. coal production halts, buyers of our coal, like Brazil, India, and Japan, are left to fill in the gaps on their own. They'll have to get the coal from somewhere else. Thanks to U.S. innovation in coal production, US coal boasts the highest quality coal in the world. It is far cleaner than coal from China. Instead of buying coal from the U.S., a country like Japan will probably turn to its neighbor China for coal. Unfortunately, China's coal is much more emission-heavy (in its mining and consumption) than U.S. coal.

We might think we're saving the planet by stopping coal production in the States, but we could be causing more harm to the *global* environment as others do a slapdash job of harnessing energy. Brazil uses a mix of its own coal and US coal, and its coal is on the dirtier end of the spectrum. Cut coal sales to Brazil, and the nation's emission profile can quickly increase.

Another example is the U.S. natural gas and oil exports to Europe and the United Kingdom. While Europe has made leaps and bounds in renewable energy, it is in a similar position as the

U.S.—most nations still need fossil fuels to keep the grid up and running. Without U.S. energy exports, many European nations, desperate for affordable energy to maintain their economies and standard of living, turn East toward countries like Russia. This wouldn't usually be a problem, except for the fact that Europe is trying to keep its distance from interactions with Russia since Russia's unprovoked invasion of Ukraine. It's difficult to slap trade sanctions on a nation if you need to trade with them.

It might seem like the U.S. can cut the use and sale of all fossil fuels, if not nationally, globally. But this is clearly untrue. The U.S. is one of the world's largest consumers and producers of energy in the world. Changes in U.S. energy policy affect global energy markets by altering the supply and demand for oil, natural gas, and coal.

While the U.S. decides on its domestic energy policy, it needs to consider the global ripple effects of its energy production.

## MOVING TOWARD THE RADICAL MIDDLE

Change in the energy sector has been slow because it is a highly complex system, and ratepayers depend on the current system for reliability and affordability. Yet, the system does need to change. Counter to what some environmentalists might claim, the sky isn't falling—however, there is a strong case for being environmental stewards. Energy is essential for a basic standard of living, but so is clean water and air. It's just that there is often a hierarchy when it comes to standard of living.

As a graduate student in Chicago, I became keenly aware of this standard-of-living hierarchy of needs and what really matters to people. In 2018, I worked during the day in Chicago's financial district, commonly called "The Loop." My work day was spent listening to energy executives discuss saving the planet and green initiatives. At night I was busy consuming classwork and the latest economic and political reports. But my commuting time, I spent just observing.

To get to the University of Chicago on the city's South Side, I'd commute on trains and buses through Chicago and watch the storefronts, the homes, and the cars change as I traveled south. Up in the north, you'd see Mercedes Benzes and BMWs. By the time I reached the University of Chicago campus on the South Side, you'd see beat-up Chevys with faded paint jobs and a variety of dinged-up Hondas.

I needed to be on campus for classes twice per week. Once, as I walked onto campus, a guy with scraggly hair, a dirty oversized winter coat, and gray-brown shoes walked by the entrance to the pristine campus. And I thought: *I bet this guy doesn't give a shit about the latest green initiative.*

At the same time, across the street, a car rolled to a stop, and a woman in nurses scrubs got out of the car, yawning. As she approached the house, a different woman emerged from the front door in a uniform. The first woman handed the keys to the other woman, who got in the car and drove off. I thought to myself: *I bet they don't care about their energy company's green promise either.*

The people who share resources to get by, wander the streets looking for shelter, or work three shifts to make rent, are usually more concerned with their day-to-day problems, than saving the environment.

Energy is vital to standard of living, but to positively affect standard of living, energy must be accessible and affordable. It's great to have renewables contributing to the power grid, but they are not enough to power the grid, nor are they affordable (yet). We need fossil fuels to literally keep the lights on.

My job today is different from back in 2018. Instead of pandering to executives, I now educate energy companies, engineers, politicians, policymakers, activists, and environmentalists on how the energy sector works. I've been deep in the industry long enough to understand how each group views themselves (favorably) and others (poorly).

The most frustrating part of my job now is helping everyone understand each other. I speak to different groups several times per week, year-round, but often, I only have an hour or two to present information and answer questions. With people like Barbara in the audience, I know I need more time.

Ergo, this book.

In the first few chapters, we'll dive deeper into energy and all its moving parts, plus dig into the nuanced knock-on effects of energy, both economically and politically. We'll look at case studies of energy companies that have been able to change quickly. We'll uncover the policies motivating energy companies to change and those producing the opposite effect. We'll better understand the heated tribalism surrounding fossil fuels and environmentalists and how working together might better serve all of our goals.

In the book's second half, we'll discuss how each group—energy folks, policymakers, and environmentalists—can better communicate and push their own agendas forward. With a better understanding of how each group thinks and operates, plus the facts, we can meet not on the ends of the spectrum but in the Radical Middle. Finally, I'll equip readers with tools for turning around a conversation that has moved from fact to opinion.

The pace of change in energy has stagnated, and everyone involved is frustrated. I've observed energy companies, policymakers, and environmentalists all blame each other rather than try to understand each other's motivations. It seems no one can comprehend why the other would push forward a particular policy or initiative. This book attempts to illuminate a path forward to collaborate for better, faster, more impactful solutions in the energy sector. We must consider all facets of the complicated energy puzzle to solve the coming energy crisis while maintaining our standard of living in the US and increasing the global standard of living. When presented with the full scope of energy's impact, most people acknowledge that energy policy moving forward must emphasize the importance of an "all of the above" set of solutions.

But before we can get into all of that, we need to understand how energy became so polarized to begin with. It starts with the villain origin story of fossil fuel companies...

# CHAPTER 2
# THE VILIFICATION OF ENERGY COMPANIES

Environmental groups and energy companies see each other with increasing hostility. A deep-seated us-versus-them mentality polarizes the discourse, pushing both sides into their respective corners. Under the unspoken rules of this distinct tribalism, even acknowledging the merits of the opposition's argument can be considered treasonous.

The increased vilification of energy companies in recent decades can be traced back to three factors. First, energy companies have been slow to admit to their flaws, and, in some cases, even tried to cover up their wrongdoings. Second, energy companies are increasingly lumped together with big corporations and assumed to share distinct "evil" traits. Finally, the environmentalist movement has been wildly successful in dominating the current narrative on energy by demonizing both legacy energy companies and the utilization of fossil fuels.

It's difficult to pinpoint exactly when energy companies began to obfuscate their practices, however, there is one major court case that molded the modern perspectives on energy companies. To be fair, the attention and action it garnered were wholly warranted. When the case caught media attention, it became clear the energy company had misbehaved for decades.

Pacific Gas & Electric (PG&E) operated a compressor station in Hinkley, California, as part of its natural gas pipeline system. To extend the life of their equipment, PG&E decided to use hexavalent chromium, a toxic but effective rust inhibitor.

Rust prevention was essential for maintaining the integrity and operation of PG&E's cooling systems. However, their disposal method involved pumping the chromium-laced wastewater into unlined ponds. With few environmental regulations to consider, PG&E was mainly concerned with their bottom line. PG&E continued to dump the toxic water into unlined ponds, which leached into the groundwater, contaminating the local water supply irreparably.[1]

The discovery of the contaminated groundwater was not immediate. PG&E's dumping of the toxic chromium occurred between 1952 and 1966, and it took decades for the implications of these actions to come to light. It wasn't until Hinkley residents began reporting unusual rates of illnesses, including cancer, that investigation began. A legal clerk was instrumental in linking these health issues to the water contamination through her own digging. The clerk found the water supply was contaminated with high levels of hexavalent chromium *and* PG&E had been aware of the potential risks and the presence of the pollutants. It was proven in court that PG&E had known since 1965 there was a "chromate problem," yet, they didn't report the contamination to the local water board until 1987.[2]

PG&E did themselves no favors as the press circled around the case. At first, PG&E attempted to downplay the risks associated with hexavalent chromium and disputed claims about its role in the community's health problems. When PG&E first reported the dumping in 1987, they claimed they only knew about the situation a week prior. More investigation revealed they knew it was an issue for several decades.

As the facts began to emerge, media coverage detailed how PG&E was not just as negligent but actively engaged in spin campaigns and misinformation. The public perceived that a large utility company knowingly and unsympathetically put public health at risk for profit. This outlook catalyzed intense backlash against PG&E and the energy sector as a whole, and was exacerbated by the personal stories of residents who suffered from serious health issues due to the contamination. As a result, environmentalists gained the upper hand and benefit of the doubt in their fight against energy companies.

With the evidence (and growing public opinion) against them, PG&E shifted its strategy. After losing $120 million in their first arbitration, they moved away from arbitration to settlement negotiations. In 1996, they agreed to a $333 million settlement—one of the largest in a direct-action lawsuit.

The public viewed PG&E's behavior with scrutiny. The extensive media coverage following the multi-million-dollar settlement impacted not only PG&E's image but also the perspectives of their investors. There were, of course, substantial financial and regulatory consequences for the company. The case became a landmark in environmental law and a lesson in corporate accountability. It fundamentally shifted the public's opinion of energy companies and perception of regulation.

During the investigation and court cases, PG&E were their own worst enemy. They feigned responsibility, which exposed a behavior pattern many saw as typical of large energy companies. The case provided the environmentalist community a gold mine of material to support the vilification of PG&E's and the energy industry by proxy.

The case eventually received renewed attention a few years later with the release of the Oscar-nominated film *Erin Brockovich* in 2000. The film dramatized the court case and brought a Hollywood spotlight to the issue. The movie was released four years after the settlement and added new fuel—no pun intended—to the anti-energy company fire. The movie's success amplified the public backlash and the modern vilification of energy companies. After all, PG&E was the villain in the film. The case is even commonly referred to today, with constant press releases updating the public on the town of Hinkley and Brockovich.

PG&E's actions were categorized as the same as other big corporations' mistakes and quickly morphed into a tool to facilitate political rhetoric. After the case, it became convenient to lump energy companies together with other villainous corporations to simplify narratives when pitching platforms to voters. By and large, energy companies had escaped this particular scrutiny up until the Hinkley, CA chromium problem. The public court case, the media storm surrounding it, and the Hollywood movie (along with regular media updates) dragged energy companies out of the dark and into the spotlight. Energy companies were now widely recognized as potential threats to public health and safety, and the general public demanded greater transparency to prevent their towns from experiencing similar ecological trauma.

The initial responses from other energy and fossil fuel companies didn't help improve the industry's collective reputation. Rather than voluntarily adopting increased transparency to prove they are doing their best, some decided to shutter themselves away. (Today, it is typical for publicly traded companies to provide sustainability reports and for private companies to withhold similar reports, since they don't have to.) Despite the critical service they provide, energy companies are typecast into a specific villain's role and, until a couple of years ago, didn't appear too bothered by it.

## PLAYING THE BAD GUY

Energy companies haven't always been vilified like they are today. Immediately after the energy shortages of the late 1970s, the energy sector was generally considered critical for the revitalization of American global economic dominance. In 1981, the *New York Times* stressed, "There is a growing realization that both conservation and increased domestic energy production are necessary and feasible...America is richer in total energy resources—oil, gas, coal, uranium, shale, synfuel technology, etc—than the Arabs."[3] Moreover, since the Industrial Revolution, the energy sector has been a driving force behind virtually all innovations impacting the standard of living. Energy companies have been at the forefront of progress, from providing the infrastructure powering new gadgets in our homes to turning gears on manufacturing equipment.

However, today, the U.S. does not typically evaluate energy within a historical context. For millennia, energy generation has been "dirty." The air in many Western 19th-century cities was thick with smoke from factories burning fuel to power manufacturing and metalworks. Wood-burning fireplaces and kerosene lamps turned wood trim in homes from light to dark. Trains pumped black fumes into the atmosphere as they crisscrossed counties and countries, transporting people and goods. To produce heat, to produce electricity or energy, there needs to be some sort of chemical reaction. And for thousands of years, the reaction has been burning fossil fuels (oil, coal, gas).

Only in modern history did electricity become the primary way of consuming energy. Instead of burning kerosene lamps to see at night or having a weekly delivery of ice, today most people flip a switch to turn on a light and own refrigerators to keep their food cold. But the electricity coursing through our homes is largely produced by fossil fuels—and few people think twice about how that electricity is generated.

Since 1975, the United States has more than doubled its electricity use. And as generative artificial intelligence evolves, as the US brings back crucial manufacturing to its shores, electricity demand will also grow exponentially.

Energy companies continued to stir up dissent for their poor environmental practices during the late 1980s and '90s. Media and public opinion cast them as the "bad guys," and energy companies responded by playing the part. Many retreated into defense, rather than proactively addressing concerns about their practices. They relied on their public relations firms, whose out-of-touch methods only served to erode public trust further.

The pattern of corporate aloofness and slow response to environmental disasters became a recurring theme with the PG&E case, but before then, other energy-sector companies were hurting the industry's reputation. The Exxon Valdez oil spill in Alaska is a prime example.

On March 24, 1989, the Exxon Valdez oil tanker struck a reef in Prince William Sound, Alaska. Roughly 11 million gallons of crude oil spilled into the ocean, making it the largest oil spill in U.S. history at the time. The media wasted no time spreading pictures of animals covered in thick black oil.

Exxon's initial response was sluggish. They underestimated the scale of the disaster and did little to contain the spill. As oil spread across 1,300 miles of coastline, images of oil-coated wildlife and ruined beaches flooded the media. Yet, public perception held that Exxon seemed more focused on minimizing its liability than the urgent cleanup effort required.

This aloof corporate response only fueled public outrage. Exxon came across as a company more concerned with its bottom line than the devastating environmental consequences of its actions. Even when Exxon finally mobilized a more extensive cleanup effort, it was seen as too little, too late. The damage to the ecosystem and Exxon's reputation was done.

The Exxon Valdez disaster solidified the public perception of fossil fuel companies as indifferent polluters. It also highlighted a troubling pattern during that time: when confronted with environmental failings these companies often retreated into a defensive crouch rather than taking proactive responsibility. Rather than quantitative transparency, they relied on PR spins. In other words, they disproportionately leaned on the subjective (which the public was not buying) over objective accountability. This deny-delay-defend playbook only reinforced the "evil corporation" narrative taking hold in the public consciousness.

Of course, not every energy company is an amoral polluter, and many have made strides to clean up their operations. The technological advances pursued by industry players have drastically improved all facets of the business, particularly those related to emissions, waste, water, and spills. In the wake of the Exxon Valdez spill, the industry significantly improved all facets of environmental stewardship, not just tanker safety and spill response.

The biggest fossil fuel and energy companies have long considered themselves indispensable to modern life. In fairness, indeed the populace is reliant on their operations. They power homes, fuel transportation, facilitate innovations in healthcare, and drive the global economy. Unfortunately, this sense of entitlement can breed a certain detachment and resistance to social and market change. When environmental concerns first arise, energy companies may brush them off, assuming their own importance will shield them from serious consequences.

As environmental awareness and demands for enhanced stewardship grew over the second half of the 20th century and into the 21st, so did the backlash against energy companies' perceived poor practices. The energy companies' defensive posture made early dialogue and collaborative problem-solving nearly impossible. For example, when BP rebranded itself as "Beyond Petroleum" in the early 2000s and touted its investments in solar and wind, it was almost immediately dismissed as greenwashing, given the company's continued focus on oil and gas. Afterwards, energy companies' consistent resistance to environmental reporting drowned out almost all of their positive efforts to improve.

Additionally, the pattern of corporate deflection from the '80s and '90s has made it easy for today's media to lump the whole industry into the evil, greedy corporation camp. Along with capitalism, energy companies have been cast as the villains standing in the way of environmental progress and change. And their lackluster PR efforts have often played right into that typecast role.

To this day, the "evil corporation" narrative continues to shape public perceptions and policy debates, often overshadowing the complexities and economic realities of the real issues at hand. While energy companies bear partial responsibility for this dynamic, they are unquestionably not the only ones responsible for the existing energy policy dysfunction, misalignment, and general misunderstanding in the market. Environmental groups, too, have increasingly focused more on assigning blame than finding practical solutions. They've often lost sight of the bigger picture by embracing their role as righteous heroes fighting against villainous corporations. Generally, both groups consciously or not, prioritize tribalism over sound economic policy.

## A HERO COMPLEX

Early environmentalists were motivated by a concern for the health of the planet and its inhabitants. They raised awareness about pollution, habitat destruction, and the consequences of industrial pollution. Initially, the movement focused on tangible, immediate goals: cleaning contaminated waterways, protecting endangered species, and preserving wild spaces. These goals resonated with most of the public. People could easily understand and appreciate the value of clean air, drinkable water, and thriving ecosystems.

In the early days of the modern environmental movement in the 1960s and 70s, activists focused on addressing specific ecological problems. They campaigned against polluted rivers that caught on fire, smog that choked major cities, and toxic waste dumped in neighborhoods. The goals were clear: clean the air, protect drinking water, and preserve natural spaces for wildlife and recreation. This approach led to significant legislative victories like the Clean Air Act, the Clean Water Act, and the Endangered Species Act. The new laws measurably improved environmental quality and public health.[4]

However, as the movement has evolved over the 21st century, it increasingly embraced more abstract, ideological messaging and political goals. Environmentalists increasingly portrayed themselves as the self-appointed guardians of the planet, fighting against the shortsighted greed of corporations—primarily fossil fuel and energy companies. This hero-villain framing has turned constructive debates into matters of moral tribalism rather than pragmatic problem-solving.[5]

Objectively, this narrative was compelling and galvanizing. It tapped into a deep human desire for clear moral lines, for good guys to cheer for and bad guys to heckle. And in many ways, it was grounded in truth—at least, initially. The environmental movement had scored important victories and drawn much-needed attention to critical issues.

But as with all good vs. evil framing, it oversimplified a complex reality. Not all corporations are mustache-twirling villains, and not all environmental causes are unambiguously righteous. The hero-villain narrative has slowly turned the debates between ecological causes and energy needs into a matter of moral and tribal affiliation rather than pursuing pragmatic solutions.

As the fight over decarbonization and net-zero targets gained mainstream attention, these goals became powerful rallying cries for the environmental movement. However, in focusing so single-mindedly on these ideals, environmentalists often lost sight of the complexities surrounding the energy transition.

Calls to "keep it in the ground" sound bold and actionable yet ignore the reality that rapidly or prematurely cutting off oil and gas would cause immense economic disruption and human suffering, especially among the global poor.[6] Environmentalists increasingly gloss over the challenges and tradeoffs inherent in a rapid energy transition. Many common consumer devices like cell phones, electric vehicles, and solar panels require precious metals, typically mined in poor African nations. The precious metals trade is also largely controlled by Chinese interests,[7] creating geopolitical tensions around the manufacturing of tech as well as mining practices.

Unfortunately, with limited oversight on mining practices, the impacts on the peoples of other nations are grave. Reports on human-rights violations of workers, devastating environmental impacts for mining regions, and civil unrest are increasing and concerning,[8] but many look the other way. In many ways, the Western environmentalist perspective stresses local (or domestic) environmental protections and neglects to consider global impacts.

Most environmentalists push for policies that sound good on paper but ignore economic realities and falter in practice. Proposed policies demand the immediate shutdown of fossil fuel infrastructure without a plan to replace the energy capacity needed to maintain basic power needs for underserved communities.[9] These unyielding stances usually backfire by alienating potential allies. In their idealistic quest, environmentalists reject incremental steps in the right direction, personifying the cliché, "Perfect is the enemy of good."

Decarbonization is a necessary goal, but it is not an end in itself. The ultimate aim is to create a sustainable energy system that meets human needs while preserving ecological health. This requires a nuanced understanding of the entire lifecycle of different energy sources. There are trade-offs—both economically and societally, locally and globally—associated with innovation in an embedded infrastructure.

None of this is to say environmentalists are wrong to push for clean energy and hold polluters accountable. Globally, aggregated emissions are on the rise and show no sign of slowing down. Improved efficiency, functional decarbonization, and the constant pursuit of scalable innovation are worthy and necessary goals. But when the movement gets trapped in the hero narrative, it can lose sight of how to translate its high-minded ideals into lasting, on-the-ground results and perpetual impacts. When environmental groups get too caught up in crusading against "evil polluters," and energy companies in taking down "tree-hugging absolutists," it becomes harder to acknowledge where energy companies are making progress in reducing emissions through efficiency and innovation.

There are instructive models of collaboration, however. Environmentalists worked with industry in Canada's oil sands to establish one of North America's largest protected boreal forests, with key wildlife corridors and wetlands.[10] The Renewable Energy Buyers Alliance in the U.S. unites major corporations and NGOs to scale up clean power procurement.[11]

The hero-villain framing pushes both sides to the extremes, demanding all-or-nothing. But, sustainable energy policies will require collaboration between environmentalists, energy companies, governments, and civil society. It will mean balancing tangible environmental goals with the practical needs of energy access, expansion, and economic security.

## VILIFICATION LEADS TO POLITICIZATION & POLARIZATION

As tensions between environmentalists and energy companies grew, politicians struggled to find common ground. Rather than seeking solutions, different parties only reinforced their opposing positions. The task of crafting energy policy became a series of partisan skirmishes, with policymakers retreating to their echo chambers instead of working together.

Lawmakers' opposing dynamic typically plays out at the state level, where energy policies often break down along predictable red/blue lines. Democrat-led states like California and New York continually push renewable energy mandates and fossil fuel restrictions.[12,13] In contrast, Republican-led states like Texas and North Dakota have doubled down on oil and gas production.[14]

These polarized approaches tend to produce suboptimal outcomes—and delay energy efficiency innovation. California's headlong rush to shutter fossil fuel plants and ramp up renewables has left its grid overly reliant on intermittent sources, leading to rolling blackouts during peak demand. Meanwhile, oil-friendly states have been slower to embrace the economic opportunities the energy transition presents.

Entrenched partisanship makes implementing balanced, context-specific policies difficult. It leads to abrupt policy shifts and regulatory uncertainty as different parties cycle into power, which deters the long-term investments needed to overhaul energy infrastructure. The past few federal administrations are a case in point The Obama administration made the Paris Agreement and Clean Power Plan centerpieces of its environmental agenda. The Trump administration promptly reversed course, withdrawing from Paris and rescinding the CPP. The Biden administration attempted to restore those policies and push through green stimulus measures, namely through the Inflation Reduction Act.

This kind of herky-jerky policymaking makes it incredibly difficult for utilities, manufacturers, and consumers to plan and adapt. It also increases the burden for investors to identify sound opportunities.

At the same time, politicization makes it harder to acknowledge areas of agreement and build on them. In reality, there is significant bipartisan support for specific clean energy policies, like funding research and development (R&D), modernizing the grid, and supporting carbon capture. There are also many local examples of fossil fuel companies, unions, and environmental groups coming together to craft just transition plans for coal communities. But these stories of common ground and collaboration get drowned out by the dominant narrative of irreconcilable division.

Overcoming this polarization won't be easy, and, like most negotiations, solutions will be complex. Policymakers, advocates, and industry groups must step out of their comfort zones and build new coalitions around shared interests. This will mean environmentalists agree to balanced solutions including an ongoing role for cleaner and more efficient fossil fuels, and energy companies agree to emissions reduction targets and practical investments in low-carbon technologies. Most of all, it will require a shift in mindset—a willingness to see the energy transition as a cooperative endeavor rather than a zero-sum struggle.

Clean energy and energy efficiency can benefit everyone by creating jobs, improving public health, strengthening energy security, and safeguarding the climate. But to get there, we must move beyond the tribalized hero/villain narrative and focus on forging pragmatic policies and unlikely alliances. We must create space for nuance and good-faith engagement and acknowledge legitimate concerns. Most importantly, we cannot cherry-pick the data to support the argument we want to win. We must allow comprehensive, objective evidence to tell us how to proceed. This steadfast allegiance to a "side" is counterproductive and a waste of energy.

## LEVERAGING WHAT WE HAVE: THE POWER OF FOSSIL FUELS

As we have seen, the vilification of energy companies has deep roots, stemming from high-profile environmental disasters, defensive corporate responses, and a growing hero-villain narrative pitting righteous environmentalists against greedy polluters. This polarization has seeped into policy, leading to partisan gridlock and suboptimal outcomes.

Energy companies, environmentalists, and policymakers have reached a mutual misunderstanding, determined to focus more on scoring ideological points than finding practical solutions. This tribalism has led each side to adopt absolutist positions and ignore the legitimate concerns and contributions of the other. But as we've seen, the reality is far more complex. The needs of the energy industry dictates there is no easy solution, and every choice will have consequences.

Energy companies are not monolithic villains, and environmentalists are not unambiguous heroes (nor are all of them crazed, tie-dyed eco-warriors). "Clean" energy is not simply a matter of flipping a switch from dirty to green. "Clean," in this case, is misleading. Decreasing emissions and optimizing for energy efficiency is a complex, decades-long process. All stakeholders must work together to balance economic, social, and ecological considerations.

One key point often overlooked in the debate is the ongoing role of fossil fuels in enabling the transition to cleaner energy technology. As we saw in Manzier's story in California, fossil fuels still provide the vast majority of global energy and will continue to do so in the foreseeable future. Rapidly cutting off any existing energy sources would cause immense economic disruption and a lower standard of living, particularly in developing countries relying on affordable power to lift people out of poverty.

At the same time, we cannot ignore the need to reduce emissions in light of the coming surge of needed power as the AI and tech revolution threaten to suck the grid dry. Instead of treating oil, gas, and coal as the enemy, we need to see them as a long-term bridge to a relatively cleaner energy future. We need to use them as efficiently and cleanly as possible while investing in the technologies and infrastructure that will improve them and result in enhanced efficiencies.

Environmentalists need to accept fossil fuels' significance in the transition and work with energy companies to make the process as clean and efficient as possible. Denying the merits of collaboration is counterproductive to most environmentalists' mission in the long term. To collaborate effectively though, energy companies will be required to mitigate their impact on the earth—from air quality to water usage to ground pollution. Moreover, energy companies and environmentalists need to include policymakers in their collaborations. Then, these policy-makers must craft regulations incentivizing desired behavior (rather than only punishments for unwanted behavior) and focus on economic realities.

There are promising models to build on. In the Permian Basin, the world's busiest oilfield, companies are experimenting with using renewable energy to power drilling operations, cutting costs and emissions. ExxonMobil has committed $3 billion to carbon capture and storage projects. Utilities across the Midwest are retiring coal plants ahead of schedule and replacing them with a mix of natural gas, wind, and solar.

The clean energy transition and responsible expansion are our time's defining challenges and opportunities. Fossil fuels, for better or worse, will be a critical part of that story. By leveraging them strategically and responsibly, we can write a happier ending where we hand the next generation a cleaner, more efficient, and more prosperous world than the one we inherited. Reliable power is not a luxury but a necessity. The lights must stay on even as we clean up the grid. Fossil fuels remain indispensable, but that doesn't mean we can't start reining in their worst impacts and investing in the alternatives. The Radical Middle is about striking a balance and bringing everyone along.

To that end, there are some strategies that do not work. We need to dissect popular but misguided solutions to understand better how to improve energy efficiency and lower emissions.

# CHAPTER 3
# US VS THEM DOESN'T WORK

The vilification of fossil fuels has led current decision making to devolve into a binary of extremes: either we rapidly abandon all fossil fuels, or we resign ourselves to a hopelessly polluted future. This two-minded framing has become the dominant paradigm in media coverage, political rhetoric, and everyday discussions. But it's a dangerous oversimplification leading to blanket regulations that typically backfire.

I constantly experience this polarized rhetoric firsthand when giving live presentations, hosting webinars, and in conversations at conferences. Recently, it was in the comments section of a webinar during a talk of mine and a subsequent email exchange.

In my presentation, I showcased data-driven arguments in a webinar about the ongoing role of fossil fuels in the energy transition. The central thesis of the presentation was to address the question of what happens to the global emissions profile if the United States decarbonizes but other fossil fuel-intensive countries do not. With live chat turned on, I did my best to answer comments and questions from viewers. The group was a mix of energy professionals, engineers, environmentalists, lawyers, and local and state-level policymakers. The questions were wide-ranging, but there was one voice that was particularly active: a university academic.

The academic had a smattering of comments throughout my presentation and I didn't have time to address all of them. I could tell by his comments he was uncomfortable with the data I presented. This is always concerning to me since I use data from sources like the U.S. Energy Information Administration (EIA), U.S. Bureau of Labor Statistics, the International Energy Agency (IEA), the Intergovernmental Panel on Climate Change (IPCC), and the World Bank. When I'm questioned on the data I use in my presentations, it's usually a warning to proceed slowly and with curiosity to understand the other person's concerns better. Chances are the question-asker is more worried the data won't align with the narrative they want to be true.

I knew the best way to connect with the academic would be to have a private video call with him where we could better understand each other's perspectives. A message thread or email exchange would likely devolve into unproductive accusations. Social media is littered with ridiculous arguments on both sides—I've found the only way to truly understand someone's perspective is to engage in a one-on-one discussion. So after the webinar ended, I emailed him to set up a time to talk. My email went like this:

*I hope all is well, and thank you for attending the webinar earlier in the week. I apologize for not being able to address all your questions, but I am happy to reconnect over a call to discuss this in further detail. At your convenience, please let me know if you have availability next week. Talk soon!*

Two hours later, his curt reply was in my inbox. He confirmed he did, indeed, have a number of questions. Instead of taking advantage of a chance to connect and clear the air (or at least learn from each other), he declined, citing our different perspectives on the Paris Agreement as resounding evidence there "is too much of a gap in facts to bridge." He then proceeded to call me "anti-climate" in his next sentence. ("Anti-climate" is not a label I subscribe to, though it is frequently applied to me by people who (typically) don't like the unbiased data on energy usage in the US.) He closed his email with: "It appears that we just have different theses overall."

In response, I expressed my disappointment that he was unwilling to connect and clarified my thoughts on the Paris Agreement in case they had been misconstrued. My argument is that the Paris Agreement is not at all legally binding since nearly 200 countries are in breach of one of their primary tenets —passing 1.5°C in 2024 has resulted in zero court proceedings and virtually zero outcry from climate enthusiasts. (More on the Paris Agreement coming up.)

Our exchange illustrates the breakdown in constructive discourse. This guy was so mired in his stance against fossil fuels he couldn't entertain a fact-based discussion about their role in enabling cleaner technologies. He considered any acknowledgment of fossil fuels' ongoing utility as a betrayal of his environmental principles.

All-or-nothing stances like these ignore basic realities. Fossil fuels still provide over 80 percent of global energy.[1] Rapidly cutting them off would cause massive economic disruption and human suffering, especially in developing nations, including countries the U.S. considers important allies. Even the most aggressive clean energy investments and policies can't change math overnight.

Maybe the academic knew a few things I don't—in fact, I'm sure he does. He has had a completely different role in the energy sector. Looking him up I discovered he worked in clean energy development. It's highly likely he could lend me insight on the nuances of coming tech I maybe haven't considered. (Hence my suggestion we have a call to discuss.) But his unwillingness to have a conversation—based on one difference of perspective—halted any movement toward practical, pragmatic solutions for efficient energy while maintaining standard of living.

As soon as he discovered I didn't fully agree with him, the academic treated me as if I were an enemy. This happens all too often in public discourse. As soon as we discover (or think) someone isn't fully aligned with our views, we lump them into the opposing extreme view. Conversation becomes stilted as we take an adversarial stance and prepare to defend our views.

Ideological rigidity, combined with tribalistic stubbornness, has become all too common, and it's a major barrier to progress —and because it's all-or-nothing (usually, "climate" versus "anti-climate"), there is no room for the kind discussions we desperately need. And, we have a variety of individuals influencing the public narrative on energy who, frankly, know almost nothing about how the broader energy spectrum works.

Any energy transition will not be a light switch that can be flipped overnight. The transition from coal to oil and from oil to natural gas didn't happen overnight. Energy transitions are multi-decade processes requiring a diverse mix of technologies and fuels, carefully phased in over time as we build out the necessary infrastructure and capabilities to meet ever-increasing demand. Yes, renewable energy will play an increasingly central role, but it's not yet ready to shoulder the entire load, especially for energy-intensive industries and developing economies. In fact, it may never be ready to shoulder such a burden.

Fossil fuels, for better or worse, will remain a significant part of the global energy system for the foreseeable future. The question is not whether we use them, but how we use them. Consider the case of natural gas (a topic I almost always mention in my presentations). Over the past decade, the shift from coal to gas in the U.S. power sector has been the single largest driver of emissions reductions.[2] Without gas as a backbone, the clean energy revolution would have been far slower.

The current us-versus-them narratives informing our discussions and policy lead us astray: they do not move the needle in terms of creating change, and often, when they do create change, it lacks foresight and nuanced understanding of how energy works domestically and internationally. Moreover, the sweeping changes enacted by both extremes often leave out key stakeholders in the energy industry.

## LEAVING OUT THE MIDDLE GUYS

When energy debates focus on all-or-nothing options, small and mid-sized energy companies are often left out of the conversation. These "Middle Guys" are disproportionately impacted by blanket regulations and one-size-fits-all policies, but they are frequently overshadowed.

Consider the challenges faced by oil and gas operators in Oklahoma's Mid-Continent (Mid-Con) basin. This region is predominantly home to many small, family-owned businesses that have been producing energy for generations. They live in the communities where they operate and have a direct stake in being responsible stewards.

But in recent years, these companies have been squeezed from all sides. Insurance costs are skyrocketing due to the perceived risks of fossil fuel operations.[3,4] Access to capital is drying up as investors and banks face pressure to divest from oil and gas. And a slew of regulations has imposed onerous compliance burdens.

The result is a severe cash crunch for the Middle Guys. Unlike large public companies that can tap debt markets or issue new equity, these smaller operators live and die by their free cash flow. Every dollar spent on insurance, interest, and compliance is a dollar not spent on upgrades, equipment maintenance, or workforce development.

Many of these operators are based in the very communities where they operate. They have a direct stake in being good environmental stewards, if only to protect their own homes and families. But when capital scarcity forces tough choices, economics often dictates doing the bare minimum to stay within the letter of the law.

This isn't to excuse bad behavior or suggest these companies should be exempt from oversight. But it does highlight the unintended consequences of a regulatory approach treating all industry players the same, regardless of size, location, or circumstance. For a small operator already running on thin margins, extreme regulations can be a death sentence because the price of their product is set by global markets beyond their control.

While smaller oil companies struggle to survive under regulations, behemoths like ExxonMobil hardly take notice. For them, even the most ambitious climate regulations are akin to a mosquito bite. Exxon has the financial resources and legal muscle to adapt and even shape the rules in their favor. In fact, they might even welcome certain regulations if they think it will hurt their competitors—aka the small and mid-sized producers.

Unable to keep up with the demands of regulations, the Middle Guys have to close, often selling for a bargain to a major energy conglomerate. As a result the market becomes more concentrated in the hands of a few major players. If a midsize or small producer doesn't sell, other consequences may be felt in the community, including increased well abandonment.

The impact of the Middle Guys closing is particularly severe in smaller markets, where local operators are crucial to the economy. When these companies fold, the surrounding communities face significant economic setbacks, losing jobs and local revenue streams. Furthermore, the responsibility for abandoned assets shifts to taxpayers, exacerbating the economic strain. The shut-down of mid-sized energy producers undermines both the environmental and economic well-being of these communities.

Some might argue this is the necessary price of progress, that these Middle Guys are collateral damage in the war against climate change. But this view ignores both the human costs and the practical realities of the energy transition. Imposing disproportionate regulatory burdens solely to damage smaller market participants, which are often key for avoiding energy monopolies, steering innovation, and providing an affordable service to locals.

The world will still need oil and gas as we shift to cleaner alternatives. Small and mid-sized operators would welcome measures to increase efficiency (thereby decreasing emissions) so as to expand their currently miniscule margins. Regulating them out of existence will only force them to live the business equivalent of paycheck-to-paycheck. And if a business is living on the margins, there is no capital left to invest in innovation (or mere maintenance).

Because current dialogue and policy is pushed to extreme ends of the ideological spectrum, blanket regulations to "fix" problems keep cropping up. But one-size-fits-all mandates leave out the Middle Guys. Too much regulatory pressure drives the small competitors out and concentrates more market power in the hands of the giants who are less incentivized to change. Targeted incentives, flexible compliance pathways, and public-private partnerships could help the Middle Guys stay afloat while still driving progress toward energy efficiency and industry transitions. The Middle Guys have an outsized stake in energy policy, and must be included in discussions on what would help them achieve energy efficiency (and decreased emissions).

Pragmatic solutions require engaging all stakeholders in an honest dialogue about how to balance economic security and environmental imperatives. And that must include the Middle Guys, the small businesses and local producers who have an outsized stake in getting this right.

Unfortunately, the polarized debate and all-or-nothing measures has led to another concerning trend: blanket divestment from fossil fuels. As environmental groups pressure investors to pull their money from oil and gas entirely, it starves even the most responsible companies of the capital needed to clean up their operations and invest in lower-carbon technologies. Ironically, it also forces the divestment proponent to lose their seat at the table along with their respective say.

## UNINTENDED CONSEQUENCES OF DIVESTMENT

On the extreme-end of the spectrum, the guiding rationale behind fossil fuel divestment is to incentivize investors to take money away from fossil fuel companies and instead funnel it into clean energy solutions. The logic goes, if investors weren't so enticed by fossil fuel companies, they would invest more in solar panel start-ups, battery storage tech, and nuclear fusion moonshots. To be clear, this stance is incorrect and counterproductive. Starving fossil fuels of capital has unintended consequences that can actually hinder the clean energy transition.

In 2024, major asset management firms like BlackRock and Vanguard have scaled back their support for environmental, social, and governance (ESG)-related shareholder proposals.[5,6] This retreat highlights the tension between the pursuit of ESG initiatives and the primary fiduciary duty of generating attractive returns on investment (ROI). For firms like these, the ability to innovate and drive positive change often depends on their capacity to maintain robust cash flow. That cash flow, in turn, funds research and development (R&D) initiatives fostering sustainable solutions. When shareholder proposals prioritize immediate ESG goals over long-term financial performance, firms might struggle to reconcile these interests with their commitment to ROI, ultimately impacting their ability to invest in the future.

One significant argument against divesting from companies not fully aligned with ESG principles is that it removes investors from critical decision-making processes. When investors divest, they effectively "lose their seat at the table," forfeiting the influence they might otherwise wield to advocate for changes from within. By remaining invested, firms like BlackRock and Vanguard can continue to push for ESG considerations in a way that aligns with financial performance goals, allowing them to shape company policies and strategies over time. This approach underscores the importance of balancing ESG objectives with the need to generate profits, as sustainable growth often depends on long-term investments that maintain a focus on innovation and profitability.

This is a recurring pattern. As capital flees the fossil fuel sector en masse, it doesn't just hurt the "bad actors"—it impacts the ability of all companies, particularly the most forward-thinking ones, to fund research and development (R&D) into cleaner technologies and more efficient practices. The divestment community tends to ignore the cross-sector ripple effects, due to the interconnection of all industries with the fossil fuel space. Divestment is a blunt instrument that often backfires, slowing rather than accelerating the energy transition.

Many forget fossil fuel companies have the scale, the engineering expertise, the infrastructure, and the political clout to drive transformative change. Many of them have seen the writing on the wall—they are keenly aware they are in danger of falling behind if they don't uncover the next big innovation first and get it to market. Cutting them off from capital only reduces their ability and incentive to be part of the solution.

Of course, not every fossil fuel company is a responsible corporate citizen. There are plenty of examples of fossil fuel interests obstructing progress and doubling down on dirty practices. Pressure from investors, civil society, and practically-minded environmentalists is necessary and appropriate to hold these companies accountable and force them to align their business models with increasing energy efficiency.

Pressure on fossil fuel companies and energy operators should be targeted, logical, and strategic, rewarding companies for taking meaningful steps to improve efficiency and reduce emissions. With savings from improving processes, fossil fuel companies can then divert capital to more improvements and R&D for innovation. Regulations should include more rewards (rather than only punishments) which will steer capital toward the most promising clean energy solutions, whether those come from startups or oil giants.

Blanket divestment from the fossil fuels industry results in missed opportunities to support projects that could make a real difference and the companies who are driving progress. Fossil fuel companies often have the most capital to play with, and so can devote more resources to energy innovation. However, if investors pull out financial support, policies levy heavy taxes, or divestment is encouraged, the biggest companies are left without capital to reinvest in innovation within their own sector. Divestment might satisfy an urge to punish "Big Oil," in the short term, but it doesn't necessarily accelerate the transition to a cleaner energy system over the long-term.

The reality is that we will need an all-hands-on-deck effort to build a clean energy future, one that includes responsible players from across the fossil fuel landscape. Starving the industry of capital only makes collective effort harder. Engaging strategically to drive innovation and accountability is the smarter path forward. The goal should be to lift up the best operators while cracking down on the true bad actors. Dangle carrots, don't only threaten with sticks.

Pundits and environmentalists like to harp on what the US is not doing in terms of clean energy, but it's also vital to understand fossil fuel usage (and emissions) on a global scale because countries use energy and emit toxic waste differently, based on their own goals, capabilities, socio-economic wherewithal, and geo-political ideology.

## GLOBAL RIPPLE EFFECTS

Americans are constantly bombarded with debates over energy policy focusing almost exclusively on reducing U.S. emissions, yet one of the greatest challenges is reducing emissions *globally*. Ideological debates not only prevent us from reaching pragmatic solutions in the U.S.—they distract us from paying attention to global energy policy.

Consider the Paris Agreement, which the academic and I disagreed on. In 2015, world leaders gathered in Paris and committed to greatly reducing emissions. Reaching their ambitious targets would mean everyone would have to reduce emissions, *together*. It was hailed as a historic breakthrough, the first time nearly every nation had signed up to a binding agreement to tackle climate change.

But there are several glaring holes in the Paris framework. Most notably, it lets developing countries off the hook, with almost no process by which to hold them accountable. China, India, and other emerging economies face much laxer requirements for reducing emissions than wealthy nations, and their commitments are often vague and conditional.

The Paris Agreement reflects a longstanding principle in climate diplomacy called "common but differentiated responsibilities." The idea behind this principle is that developed countries, which have historically emitted the most carbon, should take the lead in cutting emissions. Meanwhile, developing countries should get more leeway to prioritize economic growth and poverty alleviation. This makes a certain amount of sense when it comes to standard of living. It would be deeply unfair to demand India, where hundreds of millions still lack basic electricity access, stop burning coal tomorrow. By contrast, the average American emits over 15 tons of CO2 per year, more than three times the global average.

But this binary approach—stringent targets for the West, hands-off treatment for the rest—ignores some crucial realities. First, developing country emissions are already skyrocketing as they industrialize and lift people into the middle class. China is now the world's largest emitter, releasing more than the U.S. and EU combined, and showing no signs of slowing down or even leveling off their emissions profile.[7,8] India is third in the world for emissions and rising fast.[9,10] Even if the U.S. and the European Union dropped their emissions to zero tomorrow (currently impossible), this doesn't put a dent in the emissions of countries like China, India, and Brazil. Currently, assuming the climate is in dire need of emissions reductions, there is no path to climate stabilization that doesn't involve bending the emissions curve in the developing world.

Second, we can argue that giving developing countries a pass reduces their incentive to invest in clean technology and innovate for energy efficiency. Why adopt expensive renewables when you can keep burning cheap coal without consequences? The Paris framework doesn't do enough to make clean energy or energy efficiency attractive for emerging economies. And if you can't hit your targets, there's always the option to pull out of the agreement and carry on polluting as you like.

Of course, few people are having nuanced conversations about these types of big global agreements. International policy documents like the Paris Agreement quickly become divisive. Environmentalists hold it up as progress, and fossil fuels proponents lambast it for potentially harming trade and the global economy. Of course, this taking sides becomes problematic because then neither group looks too closely at what the agreement says or how it will work in practice. And this trickles down to everyday conversations, where energy companies feel they must align against such environmentally-minded agreements, and environmentalists must side with it, even if both sides have more nuanced perspectives on it.

As I clarified to the academic, my concern with the Paris Agreement is that it has no teeth. This quickly devolved into taking sides. He became obstinate, citing my criticism of the Paris Agreement to mean I don't care about the environment at all. In truth, my concern is the Paris Agreement will become a repeat of the Kyoto Protocol.

Adopted in 1997,[11] The Kyoto Protocol required countries to set emission reduction targets and face fines if they didn't meet their goals. However, countries could simply leave the agreement and face no penalties if they fail to meet reduction targets...which is exactly what happened to Canada.

Canada made lofty promises to reduce its emissions, however, when it became clear they would not be able to follow through, the choice became obvious. If Canada stayed a part of the Kyoto Protocol they would face $14 billion in fines.[12] Such penalties would devastate any country's economy. Canada withdrew from the Kyoto Protocol, as would any country. Policymakers display a distinct tendency to promote climate-related goals until it is time to face harsh realities. Conveniently, once reality sets in, climate priorities magically change.

If ideological pontificating was out of the way in our discussions with one another, we could have a more nuanced conversation about international energy issues. This is crucial because the limitations of the Paris Agreement point to a broader issue: many climate solutions are designed by and for the West, without adequate consideration of developing world realities. Renewable portfolio standards, carbon taxes, electric vehicle mandates—these may work well in California or Germany, but they falter in countries where energy affordability and access are more pressing concerns than decarbonization. In fact, the developing world cares very little about decarbonization—their primary focus is to enhance standard of living within their own boundary lines.

But because of astringent polarization in the energy industry, we don't have these conversations—and they don't ripple out to the masses. In the West we might celebrate every new Tesla Model S, Kia Ioniq 5, and Chevy Bolt EUV on the road as a climate win. But the minerals needed for those EV batteries—cobalt, lithium, nickel—are often mined in developing countries with weak environmental and labor standards.

The Democratic Republic of Congo supplies over 70 percent of the world's cobalt, often mined by hand in dangerous conditions.[13] Indigenous communities in Chile's Atacama Desert are battling lithium miners for scarce water resources.[14] Indonesia is burning rainforests to make way for nickel mines.[15] The clean energy revolution, it turns out, has a very dirty underbelly. Even more concerning is the environmental movement's continual refusal to acknowledge this critical portion of the narrative.

Similar stories play out with other "green" technologies. Solar panels and wind turbines require rare earth elements, mostly mined in China under lax environmental rules.[16] Hydropower dams can disrupt river ecosystems and displace communities.[17] Biofuels compete with food production and accelerate deforestation.[18]

The false dichotomy between fossil fuels and renewables is not helpful here. Developing countries will need both in the near term - the question is how to rapidly shift the balance. Blanket fossil fuel bans or divestment campaigns risk leaving them in the lurch. Instead, we need targeted policies making clean energy the most attractive option, while still ensuring reliable, affordable access to the masses.

It seems in the hubbub of ideological debates, those arguing for all-or-nothing policies forget to factor in the far-reaching effects as well as the problem of global emissions. But there's another nuance to the energy sector many national debates leave out: the small to mid-sized energy companies who are disproportionately affected by extreme-ends-of-the-spectrum policies.

Ultimately, overcoming the polarization and tribalism currently dominating the energy debate will require a fundamental shift in mindset. All sides must climb out of their trenches and recognize their shared stake in a sustainable, prosperous, and just future. We must remain intellectually honest with the historical dynamic—extreme-ends policy haven't worked and currently aren't working in the great energy transition. We need to meet somewhere in the middle.

## EXTREME COLLABORATION

The polarization and ideological tribalism characterizing much of the current energy debate create significant barriers to the kind of cross-sector collaboration needed for effective decarbonization. As the earlier sections of this chapter have shown, when proposed solutions are even adjacent to fossil fuels, they are often quickly dismissed due to entrenched ideological differences. Similarly, when any kind of regulations or caps are suggested, they are immediately rejected by energy companies who see all environmentalists as enemies. Instead of pragmatic progress, most measures are made to cater to the extreme ends of the spectrum and thus opportunities to improve the energy sector are missed.

Blanket divestment impedes fossil fuel companies' ability to improve efficiencies and scale functional technologies. Overly strict environmental regulations can drive small and mid-sized energy companies out of business, allowing big energy corporations, who are more immune to regulations in the first place, to take over a greater portion of the sector. Intense debate on domestic policy also distracts from creating or incentivizing needed changes in global energy policy.

Moreover, the narrow framing of the debate often ignores the complex global ripple effects of energy policies. Decisions made in Western economies can have unintended consequences in developing nations. Abrupt shifts away from fossil fuels can disrupt energy access and affordability for billions of people still reliant on these sources. Surging demand for materials used in clean energy technologies can fuel environmental destruction and exploitative labor practices in source countries. Emissions reductions in one region can be offset by increases elsewhere if policies don't account for international emissions.

The false dichotomy between fossil fuels and renewables, the unintended consequences of one-size-fits-all measures, and the discounting of key stakeholders like the "Middle Guys" all illustrate how basing solutions on current ideological narratives often does more harm than good. The complex interweaving of economic, social and environmental effects, both domestically and internationally, gets lost in the noise of polarized rhetoric.

Moving forward will require a shift in mindset from all involved parties. Environmentalists must acknowledge the ongoing role of fossil fuels in the transition and work towards making that role as clean and efficient as possible. Energy companies, in turn, need to prioritize efficiency improvements and environmental impact management. Policymakers have the task of developing nuanced, adaptive frameworks balancing economic realities with the need for clear and stable regulations. We need to abandon extreme-ends of the spectrum and meet in the middle.

The next chapter will explore what this new paradigm, the Radical Middle, might look like in practice. How do we foster the kind of cross-sector collaboration and integrative thinking essential for progress? What new models of governance, investment, and public engagement are needed? How do we balance the competing imperatives of fewer emissions, economic development, energy access, and environmental justice? By wrestling with these thorny questions, we can begin to chart a path beyond polarization and toward a more real and hopeful energy future.

# CHAPTER 4
# THE RADICAL MIDDLE MINDSET

In the world of energy, progress has been stymied by entrenched ideological positions. Setting aside preconceptions, leading with curiosity, and seeking to understand different perspectives before passing judgment.

In the energy space, energy industry stakeholders, environmentalists, and policymakers are compelled to take extreme views (for a variety of reasons we will cover later). These three groups dominate the sphere and often talk past each other, more interested in scoring points than finding practical solutions. It's so pervasive, any viewpoints in the "middle" are so rare as to be radical. But "radical" compromise is what will need to happen if we want to see progress in the energy sector. To break through these normalized extremes, we need a shift towards the Radical Middle.

The only way we can converge to the Radical Middle is to first change our mindsets. At the extreme ends of the spectrum, judgmental attitudes and unwillingness to consider others' perspectives keep us from finding pragmatic, actionable solutions. We can meet in the Radical Middle when we adopt the Radical Middle Mindset: an attitude of curiosity; a focus on energy efficiency; and a willingness to constantly adapt (to new information, environmental changes, and global forces).

The first change everyone needs to make is to shift their mindset from a position of judgment to curiosity. A scene in the hit TV show *Ted Lasso*, encapsulates the power of curiosity over judgment.[1] The show follows the eponymous character, a collegiate American football coach, as he adapts to leading a British soccer team. In the relevant scene, Ted challenges his antagonist Rupert to a game of darts. Rupert, the former team owner, is a wealthy, arrogant man who looks down on Ted's folksy optimism and unconventional coaching methods. Rupert assumes that Ted, with his Southern drawl and aw-shucks demeanor, will be an easy opponent to defeat.

The scene unfolds in a pub, with the team's current owner Rebecca and other patrons looking on. Rupert and Ted compete, with Rupert racking up an impressive score. On Ted's final throw, Rupert smirks at his opponent, confident in his impending victory. Ted needs two triple-twenties and a bullseye to win—a difficult, if not impossible, hand to throw.

Ted steps up, and remarks, "You know, Rupert, guys have underestimated me my entire life. And for years, I never understood why. It used to bother me. But then one day, I was driving my little boy to school, and I saw this quote by Walt Whitman... It said, 'Be curious, not judgmental.' I like that."

Ted throws his first dart. The dart hits triple twenty.

He speaks again, the whole bar listening in: "All them fellas that used to belittle me, not a single one of them were curious. They thought they had everything all figured out, so they judged everything and they judged everyone...if they were curious, they would've asked questions...Questions like, 'Have you played a lot of darts, Ted?'"

Ted throws another triple twenty. The patrons of the bar whoop in excitement. Rupert's cocky smile falters.

"To which I would've answered, 'Yes sir. Every Sunday

afternoon at a sports bar with my father, from age ten to sixteen, when he passed away."

Ted throws his last dart. It's a bullseye. The crowd erupts in cheers, while Rupert and his wife promptly exit the bar in defeat.

Ted's victory in the dart game is more than just a satisfying comeuppance for an arrogant villain. It's a reminder of the consequences of making assumptions and the power of curiosity over judgment. It also represents a smug elitism implying wealth is directly correlated to competence and knowledge. Rupert judged Ted and assumed Ted would be an inferior dart player. Had Rupert shown more curiosity, perhaps he wouldn't have entered into a match he was doomed to lose from the start.

This scene encapsulates a key theme running throughout *Ted Lasso*: the power of an open and curious mindset. Ted consistently defies expectations and achieves success, not because he's the most knowledgeable or experienced coach, but because he approaches every person and situation with genuine curiosity. He seeks to understand rather than judge. Although he is considered a "simpleton," his approach to problem solving is highly effective: understand first, then apply logic and pragmatism.

This mindset is in stark contrast to the polarization and tribalism characterizing the energy discourse. The energy debate in particular has become deeply entrenched, with environmentalists, energy and fossil fuel companies, and policymakers often talking past each other rather than engaging in good-faith dialogue. Each side has its own set of assumptions and judgments about the others, which prevent collaborative problem-solving.

But as Ted Lasso shows us, if we don't remain curious, we risk losing out—especially on complex issues. If we better understand each parties' skills and motives, we can take pragmatic risks. Had Rupert been curious, he may have better wagered in his bet against Ted, or chose a different game altogether. Instead of slinking off in a humbling defeat, he may have won or lost with more grace.

Shifting to the Radical Middle Mindset won't be easy. Most people are so used to judging others, they don't see it as judging. They immediately assume they are correct because they seek opinions which confirm what they want to hear. To move toward the Radical Middle, we must first adopt a mindset of curiosity. We must "Be curious, not judgemental." If we want to create the conditions for the kind of change in the energy sector everyone says they want, everyone will need to pause before firing back in conversations, emails, and online.

There is no single solution to energy. Each decision and policy represents a web of trade-offs, all of which—including renewables—require balancing externalities. Adopting the curiosity piece of the Radical Middle Mindset is particularly crucial when it comes to weighing the competing priorities and trade-offs inherent in global energy expansion. It is undeniable that each respective energy source, fossil fuels or renewables, will have trade-offs, locally, domestically, and internationally. But the tradeoffs can be efficiently managed and negative externalities can be mitigated if we remain curious as we try to balance all the competing needs in the energy sector.

## THE RADICAL MIDDLE BALANCING ACT

The Radical Middle Mindset recognizes change in the energy sector will involve tradeoffs between economic, environmental and social objectives. It's not simply a matter of either/or choices, but of finding ways to advance multiple priorities simultaneously through both/and solutions. It might seem a tall order, but it can be done—there's even precedence.

A prime example of this balancing act in practice is the LNG Canada export terminal currently under construction in British Columbia. A joint venture between Shell, Petronas, PetroChina,

Mitsubishi and KOGAS, this massive project aims to ship Canadian natural gas to Asian markets. One major goal—enabling coal-to-gas switching in the Eastern Hemisphere—could significantly reduce global emissions.

Crafting the project timeline and securing buy-in from all stakeholders was a complex undertaking. The consortium had to navigate competing interests, from the economic imperatives of the companies to the environmental concerns of local communities and advocacy groups. This required patience, flexibility and a willingness to understand different perspectives—that is, an attitude of curiosity was necessary in coming together and working on this project.

Engaging with the twenty First Nations (i.e., indigenous tribes who are neither Inuit nor Métis), along the pipeline route was central to the project's success. The proponents took the time to understand these communities' unique needs and priorities, laying the groundwork for agreements that provide jobs, revenue sharing, and environmental oversight for the Nations. This curiosity-driven approach enabled the project to move forward while delivering tangible benefits to Indigenous communities.

Reaching these agreements could not have been easy. There were undoubtedly moments of tension as different parties sought to protect their interests. But by remaining committed to the process, staying curious, and focusing on areas of common ground, the proponents and First Nations were able to find a balanced path forward. The resulting partnerships are a testament to the power of good-faith engagement and the potential for fossil fuel projects to deliver tangible benefits for Indigenous communities.

Similarly, the decision to power the liquefaction of the natural gas with renewable hydropower rather than fossil fuel turbines was a conscious decision to favor long-term environmental and social value. It also necessitated close coordination with local utilities and regulators to ensure a reliable renewable energy supply.

From a climate perspective, the most significant feature of LNG Canada is its potential to displace coal in Asian power markets. Most Western societies only look at themselves for decreased emissions efforts. However, as discussed earlier, the majority of global pollution is coming from developing nations in the Eastern hemisphere. Canadian gas could serve as a lower-carbon alternative for many Asian nations, bridging the gap between current fossil fuel dependence and a clean energy future.

This project was not an easy sell at first, particularly for those who saw any fossil fuel development as incompatible with climate action. But by taking the time to walk through the numbers and build the fact base, the proponents were able to gradually win over key stakeholders, from government officials to moderate environmental groups. They demonstrated that a pragmatic, context-specific approach could deliver real emissions reductions in the near term while supporting a longer-term shift to renewables.

What made this balancing act possible was a fundamental shift in mindset from the traditional adversarial approach to energy development. Instead of viewing First Nations, environmental groups and local communities as obstacles to be overcome, the companies spearheading the project engaged them as partners and stakeholders in finding mutually beneficial solutions.

There are still those who oppose LNG development on principle, just as there are those who feel it doesn't go far enough in prioritizing economic growth. But by striving to balance multiple objectives and engaging in good-faith collaboration, the LNG Canada partners were able to find a middle path delivering real progress on key fronts.

The LNG Canada story, for all its unique complexities, offers a hopeful glimpse of what this paradigm shift could look like in practice. To navigate the complex challenges of the energy transition, we need to embrace the ethos of the Radical Middle Mindset, seeking to understand different perspectives before taking a position. It means accepting there will be tradeoffs and compromise, and thus staying focused on common goals can aid in finding creative solutions balancing multiple objectives.

Not every project can replicate the specific outcomes achieved by LNG Canada. The unique circumstances of each development, from the geographies to the market conditions, will shape the arch of what's possible. Using hydroelectric power for a semiconductor plant in Arizona is not feasible—there's not enough water. Building solar farms in the upper Northeast doesn't make a lot of sense. There are not one-size-fits-all solutions but the principle of seeking a balanced multi-stakeholder engagement is widely applicable.

The Radical Middle Mindset and approach doesn't eliminate conflict and controversy around the project. Instead it offers a balanced set of solutions efficiently addressing conflict and controversy. A curious outlook will help every stakeholder in the energy transition recognize there are tradeoffs. Maintaining a Radical Middle Mindset we can resist the urge to become enraged by ideas that don't fully meet our expectations, and instead engage with others to drive pragmatic and long-lasting solutions.

However, there is one goal that can unite everyone and draw us all into the Radical Middle…

## PRIORITIZE ENERGY EFFICIENCY

An efficiency-first approach aligns the interests of diverse stakeholders who are often at odds. Environmentalists recognize that the cheapest and cleanest unit of energy is the one we don't use in the first place. For fossil fuel companies, efficiency improvements extend the lifespan of their resource base and help maintain competitiveness and margin in the face of carbon constraints. Utilities see efficiency as a cost-effective way to increase capital resources and use the increased gains to invest in further improvements.

A key component of the Radical Middle approach is prioritizing and incentivizing energy efficiency, functional innovation, and scalability. While reducing carbon emissions is a long-term shared goal, a narrow focus on decarbonization targets can lead to policies that increase energy prices. In contrast, a strategy centered on improving energy efficiency—i.e., getting more output from each unit of energy input—offers a pragmatic path to advance multiple objectives simultaneously. Enhancing efficiency is the fastest way to reduce emissions and pollution across the economy.

For policymakers and ratepayers, the economic case for energy efficiency is compelling. In one initiative, energy efficiency programs deliver $4.32 in benefits for every $1 invested, in the form of lower energy bills, reduced fuel costs, and avoided infrastructure build-out.[2] These savings are especially important for low-income households who spend a disproportionate share of their income on energy.

Consider the state of Virginia and their push for energy efficiency. Energy demand is growing rapidly, driven by an influx of data centers and other energy-intensive industries.[3] At the same time, the state has committed to ambitious environmental targets, creating a tension between economic development and environmental goals.

To navigate this challenge, Virginia has prioritized energy efficiency as a core solution. In 2020, the state passed the Virginia Clean Economy Act,[4] which establishes a mandatory energy efficiency resource standard. The law requires utilities to achieve incremental annual savings of 5-percent by 2025, equating to billions of dollars in avoided costs for ratepayers. It also directs regulators to develop new efficiency programs targeted at low-income customers, helping maintain a manageable set of costs.

Gains in efficiency reduce the need for and pressure on new renewable "green" tech. This is important because as of now most renewable energy tech is expensive to install and does not generate the output needed to keep up with current and predicted consumption levels. Consider the case of Eliot Manzier and CAISO in California, where it is clear renewable energy is not enough to keep up with demand. Or a state like Virginia, which is slated to increase the number of energy-hungry data centers. Virginia would have to spend billions on renewable tech to keep up with demand and likely pass along those costs to ratepayers. As the U.S.—along with the rest of the globe—becomes a more digital world, we will be using increased amounts of energy for critical infrastructure like data centers and chip manufacturing. Incentivizing energy efficiency balances economic, social and environmental imperatives regardless of location. Hopefully, at some point in the future renewable-produced energy will supply enough power to serve all our needs, but this is not the current state—and environmentalists and politicians must acknowledge it.

Another example of where incentives for energy efficiency could have a major impact is in the natural gas space. Currently in the U.S. there is a methane tax punishing natural gas producers for usages and leaks and makes little distinction between production size. Yet, natural gas is a critical energy source facilitating actual decarbonization. The stringent nature of the tax could force some facilities (likely the Middle Guys) to shut down production if they cannot comply within the given time frames, leading to potential disruptions in natural gas supply and increased energy costs.[5]

Of particular note is the penalty imposed for methane leaks. This penalty is redundant and particularly burdensome because natural gas producers are already economically incentivized to prevent methane leaks. A leak is a sign of wasted product—companies that figure out how to prevent or harness leaks reap more rewards in having more product to sell. Instead of punishing methane producers, policymakers could design regulations to reward for efficiency improvements and emissions reductions. (This might involve tax credits for implementing best practices like leak detection and repair programs, upgrading to lower-emitting equipment, or deploying methane capture infrastructure.)

Under such a system, operators would have a direct financial incentive to minimize methane waste and maximize the amount of gas brought to market. The most efficient producers would be rewarded, while laggards would pay more. This would also free up capital to drive innovation and continuous improvement across the industry. This approach would greatly benefit those Middle Guys who are often operating at cost and might otherwise struggle to absorb the financial burden of a methane tax.

The Radical Middle sees energy efficiency as the foundation for a more collaborative and solutions-oriented approach to the energy transition. Energy efficiency becomes a common goal, advantageous for all stakeholders. It also allows for more nuanced discussions focusing on pragmatic solutions and helps everyone avoid becoming entangled in ideological debates.

We need a "back-to-basics" mentality to dominate the global energy narrative—one rooted in science and economics. Rhetoric and political dysfunction have created a regression in energy policy. An efficiency-first mindset embodies the pragmatism and adaptability essential for navigating an uncertain energy future. As technologies evolve and market conditions shift, an efficiency focus provides a North Star to guide decision-making. It prioritizes continuous improvement and optimization across all domains as new information and tech becomes available.

## THE RADICAL MIDDLE ADAPTS

In addition to a curious mindset and a focus on energy efficiency, the Radical Middle Mindset embraces flexibility and course-correction. The Radical Middle approach recognizes there is no one-size-fits-all solution to the challenges of energy policy. Moreover, as a part of adaptability, strategies need to evolve over time based on changing technological, economic, and political realities.

Part of this adaptive nature is recognizing different regions and communities will require context-specific solutions over time. A policy that works one year to incentivize energy efficiency will inevitably evolve and not work in light of new technology or economic reality. And just as a policy that works in Oregon might not benefit the state of Texas, successful domestic policy might not work well abroad. It's more work to come up with nuanced answers, but it's necessary to create marked change in global energy sectors. The Radical Middle Mindset understands there will be constant policy tweaking based on the specific needs of geographic regions, states, and communities. In short, we cannot ignore the regional geological factors when fashioning effective energy policy.

As an example of varied needs, consider what a shift toward energy efficiency and reduced emissions might look like in West Virginia. West Virginia is a historically coal-abundant and dependent state, with 89 percent of electricity generation coming from coal.[6] The state faces a challenging transition with mounting regulatory pressure to transition away from coal.

However, the economics of coal mining and power generation have a long history in West Virginia—and it's been a huge economic benefit for the state in the past. Rather than halting coal mining altogether, West Virginia could seek to increase exports of its coal to other, more coal-dependent nations. After all, the coal mined in West Virginia (and southeastern Pennsylvania) is of a cleaner variety than used in big coal-consuming nations like China and India. Coal may not be appropriate for a wealthier county in the United States, but other parts of the developing world rely on coal.

In this context, a Radical Middle strategy for West Virginia would prioritize measures to ease the transition for coal communities and to ensure costs to switching to cleaner options like natural gas (of which is also found in relative abundance in West Virginia) will not cost the average West Virginian more on their monthly energy bill. (As one of the poorest states in the country, this should be of concern for any energy-industry stakeholder—if you charge too much, no one will pay.)

However, several decades out, West Virginia might need to adapt again. China and India may have transitioned away from coal. Like all other socio-economic areas, West Virginia will need to consider what the global future might look like, using capital from the proverbial harvest to prepare for the famine. Energy producers might consider using excess capital to invest in natural gas infrastructure.

In comparison, a state like New Mexico will take a different path to increased energy efficiency. New Mexico has transitioned away from coal, and indeed has cut its coal usage in half since 2012, and 90 percent since 2002.[7] To account for the difference, New Mexico has invested heavily in wind and solar. The topography of New Mexico, particularly the high plains in the eastern region and the abundantly sunny weather, have aided the transition to renewables, which now account for 35 percent of electricity generation.

Compare New Mexico's natural assets to West Virginia and it's obvious West Virginia will eventually require a different approach to maximizing energy efficiency. West Virginia has a seasonable climate where sunny days are not guaranteed. Combine this with a topography dominated by mountains, which makes wind farms difficult to build. Neither West Virginia's weather nor topography is conducive to ramping up wind and solar efforts. If West Virginia wants to decrease its emissions profile while increasing energy efficiency, it will have to look to natural gas and nuclear power. This isn't complicated math—the proposed policy states the obvious. And indeed in West Virginia currently, renewables (wind and hydroelectric) account for only around 7 percent of the state's energy.

New Mexico will need to adapt in the future, because the only constant is change. Whether its dwindling water resources or rising temperatures across the state, there will be coming changes. Ratepayers may pull more power and water from municipal supplies, burdening the system on the hottest of weeks. This puts extra pressure on the grid. Although the wind farms are promising, they need a particular kind of geography to be most effective, and New Mexico's geography is quite varied across the state—it is not all high plains.

Across all of these contexts and more, it's clear blanket policies and political rhetoric don't work everywhere, particularly when requiring immediate shifts to renewables. Policy frameworks need to be responsive to the slow and quick shifts in energy technology, geopolitics, and the needs of ratepayers. This requires a shift away from static, prescriptive approaches towards more adaptive, outcomes-based frameworks prioritizing energy efficiency.

One example where the Radical Middle Mindset is missing (and which I still struggle to understand) is the planned construction of a major new semiconductor fabrication plant in Arizona. Semiconductors are a key enabling technology for the clean energy future, critical for everything from EVs to smart grids. With the passage of the federal CHIPS Act, there is a concerted push to reshore chip manufacturing capacity in the U.S. Yet, chip manufacturing is incredibly energy and water intensive.

Arizona, with its skilled workforce and business-friendly climate, has become a hub for new fab projects. But there's a problem: semiconductor manufacturing is incredibly water-intensive, and Arizona is in the midst of a historic megadrought. Fab projects also have substantial energy demands to power the delicate machinery and keep the facilities cooled to the necessary temperatures.

In the short-term, it might look promising to situate the semiconductor plant in Arizona. It has abundant land, a favorable regulatory environment, and proximity to key markets—plus it's offering significant tax incentives. But a more long-term analysis reveals the significant water and energy constraints that could undermine the long-term viability of this infrastructure. Certainly, water costs will become prohibitive, as the plant competes with residents and established businesses (including agriculture) for supply. The upstart costs might be lower, but the running operational costs may well be astronomical in ten years. The Radical Middle approach to finding a home for the semiconductor would mean a reassessment of the assumptions and judgments in the project for the short and long term.

To tackle the challenges of change in the energy industry, a Radical Middle Mindset is essential. The mindset is a practical approach encouraging curiosity, adaptability, and focuses on common goals like energy efficiency and affordability. Most importantly, it emphasizes finding collaborative solutions that achieve multiple goals at the same time.

Moving to the Radical Middle requires a fundamental shift in how we approach the energy transition challenge—and the first step is adopting a Radical Middle Mindset. This means setting aside preconceived notions and leading with curiosity—asking questions to better understand the incentives, motivations and constraints of different stakeholders before passing judgment. Only by truly grappling with the nuances and tradeoffs of the real world can we hope to craft durable solutions.

Ultimately, meeting in the Radical Middle Mindset sets the stage for having constructive conversations resulting in pragmatic answers to the challenges faced by the energy sector, environmentalists, and policy makers. And it's about embracing the hard work of collaboration and compromise necessary to tackle a challenge as complex as global energy policy. To that end, we need to remain curious, focus on energy efficiency, and think adaptively. To help us get there though, we'll need to draw others toward the Radical Middle with us. One key to gathering in the Radical Middle is asking better questions…

# CHAPTER 5
# MEETING IN THE RADICAL MIDDLE

A Radical Middle Mindset is the precursor to creating change in the energy sector: next we need to bring others into the Radical Middle. This is particularly challenging because most people are so accustomed to defending their own positions that our communication styles have become ineffective for encouraging collaboration. We're so used to having heated arguments, our conversations can quickly escalate. In the current climate, a dialogue can easily go from a rage-level of zero to 100 in the matter of minutes.

Of course, there will be resistance to meeting in the Radical Middle because the way we typically communicate with each other doesn't work to collaborate. Unfortunately, we cannot expect others to change their communications styles. Fortunately, with a Radical Middle Mindset, we can change our own —and it only takes one person changing to interrupt a pattern.

I notice this phenomenon all the time in my work when speaking on panels, answering Q&As, and fielding emails from all sorts of energy stakeholders. But a routine conversation with my son was the catalyst for realizing how difficult these conversations must be for other people.

"Alright, buddy," I said, walking into the living room. My

12-year-old son was playing his cherished Fortnite on what is supposed to be the family TV. He paused his game and turned around. He knew what I was going to say. It was 9:45pm on a Wednesday. His younger brother was already in bed.

"But Dad, I'm not tired yet!" he whined.

"It's a school night," I reminded him. "You need to be in bed by 10. That's the rule."

"None of my friends have to go to bed this early," he protested.

"And if your friends all jumped off a bridge, would you do that too?" I asked, annoyed to be arguing with him, yet again. I was doubly annoyed to be echoing the same line my parents used on me.

"Maybe I would!" my son shot back.

"I don't care what your friends do!" I started. "These are *our* family rules. You're going to bed now, end of discussion."

"But—"

"Your phone is mine for two days if you don't get your butt off to bed right now." Pulling the phone card was always a last-chance measure, but he needed to know I was serious.

"This is so stupid! You're just trying to control me!" he shouted, getting up off the couch, and stomping off to his room.

We'd had this conversation plenty of times before. I gave a gentle reminder of the rules, and he pushed back. I would then force him to comply, usually threatening punishment. It never made me feel good to watch him storm off, and it certainly didn't result in goodwill toward each other.

That night though, I recognized the conversation as similar to interactions between two energy stakeholders butt heads with no resolution. Yes, arguing with a tween had the same ring to it as two adults squabbling over the energy industry. Each one attempts to force their will on the other, using language and arguments that don't resonate, and sometimes even hurling verbal character attacks. Instead of being closer to pragmatic solutions, by the end of the discussion or debate stakeholders seem evermore entrenched in their extreme end of the spectrum.

From engineers and environmentalists to policy makers and fossil fuel company CEOs, everyone can struggle to have productive conversations, even when they enter into a dialogue with good intentions to be curious. Curiosity is a necessary mindset to even enter into possibly productive conversations. However, a Radical Middle Mindset crumbles if we aren't mentally prepared for conversational attacks and ready to retaliate with de-escalating follow-up questions.

In an interaction, if we feel threatened, dismissed or coerced, our natural response is to become defensive and close ourselves off from considering other viewpoints. Think about the last time you were in a heated disagreement with someone. Maybe it was a colleague who shot down your idea in a meeting, or a relative who made a snide comment about your lifestyle choices over dinner. Or maybe your son refused to follow simple family rules. Chances are your immediate reaction was not to become curious about their perspective. More likely, you felt a surge of anger, hurt or frustration. You may have jumped to conclusions about their motives, or rushed to defend your own position.

Meeting in the Radical Middle will take more than the Mindset—we need to be ready for and anticipate resistance and attacks from our conversational partners. And we need to lead with curiosity when responding, which will help us avoid retaliating, and instead lead us to asking better questions. Questions like:

- "What are your key considerations when making decisions about future investments and technologies?"
- "Help me understand your biggest priorities when it comes to the energy transition. What matters most to you?"
- "What would a win-win scenario look like from your perspective? Where do you see potential for alignment?"

Questions like these crack open the door to productive dialogue. Initially, the person on the opposite end might give you a look of bewilderment. But good questions decrease the chances the other party tunes you out completely. When we resist hurling out knee-jerk statements and accusatory questions, we remain in the Radical Middle and invite others to join us. I've discovered the reason most conversations devolve into ideological standoffs is because we charge into discussions with battering rams that only trigger a common communication barrier known as psychological reactance.

## CONVERSATION KILLER: PSYCHOLOGICAL REACTANCE

Many people are unfamiliar with the term "psychological reactance," but once explained, anyone can identify the phenomenon. Reactance theory posits that when people feel their autonomy (i.e., the freedom to act or think how they choose) is being threatened, they become motivated to regain that freedom. This often looks like resistance to persuasion, increased desire to do the opposite of what's requested, and even hostility towards the source of the perceived threat to autonomy. Psychological reactance is incredibly common and a major barrier on the path to productive dialogue between energy stakeholders.

In simpler terms, psychological reactance explains the behavior of every fight with a stereotypical teenager. It's obvious psychological reactance is to blame for the increasing tension in the argument between my son and me. He felt his freedom to do what he wants was threatened, and he lashed out, acting more contrary than he might actually feel. Of course, like an idiot, I played right into his reactance with my own reactance when I accused him of being an unthinking follower among his friends (which was unrelated to my request), and then using an ultimatum to force him to comply.

We see this dynamic play out constantly in the polarized energy debate described in earlier chapters. When environmentalists demand fossil fuel companies immediately halt all drilling and divest from oil and gas, those companies perceive it as a threat to their freedom. Taking the option to drill (for fossil fuels) off the table indefinitely could hurt a company's ability to survive in the industry. Similarly, when fossil fuel advocates dismiss renewable energy as a niche boondoggle, renewable start-up founders feel their life's work and values are being trivialized. In response, each camp often doubles-down on their beliefs and may even lash out with accusatory press release statements, marketing campaigns, and the like.

When reactance is triggered in a conversation, each side digs in their heels, and productive conversation grinds to a halt. This is because psychological reactance makes us retreat toward our ideologies and deepens the us-versus-them mentality. Each side focuses on proving their point rather than on finding common ground. And reactance isn't only triggered during perceived threats to freedom.

More research on psychological reactance has found attacks on expertise and trustworthiness arouse reactance.[1] If you're trying to create rapport, dismissing someone's credentials or disparaging the validity of their data will backfire. It might sound obvious and trivial, but, if I had a dollar for every time energy stakeholders questioned each other's facts, experience, or general intelligence, I could retire right now. On occasion, it might be entertaining to watch, however, it never results in solutions.

If we want to meet in the Radical Middle, we need to be mindful of psychological reactance, both in ourselves and triggering it in others. We can reduce our own reactance by realizing attacks are inevitable, especially in conversation about energy. People will disparage data sets, moral character, viewpoints, and expertise. During all attacks, we can resist retaliation by leaning on our curious mindset. What made the other person lash out at us? (This is not usually a question to ask out loud—that would be weird and awkward—but rather an internal question helping us pause and think instead of instantly retaliating. We'll get to what questions to ask momentarily.)

Skillful question-asking is perhaps the most powerful tool for defusing reactance and building bridges of understanding. It often turns out to be an effective tool for preventing outlandish claims and cutting to practical matters. Well-crafted questions can completely shift the dynamics of stuck conversations, opening up unexpected pathways to progress in the energy transition. By wielding questions strategically, the curious can invite others to let go of rigid positions and join the search for win-win solutions.

## THE POWER OF (GOOD) QUESTIONS

Most of us have been asked bad questions before. Some "bad" question examples are:

- What were you thinking??
- Did you ever once pause to consider...??
- How can you just mindlessly follow/trust...??

Often we know it's a bad question because there is a

presumption in the question. For example, the question "Did you ever once pause to consider...??" implies the other person didn't make any considerations in the first place. Similarly, "How can you blindly trust/follow...??" implies the answerer hasn't thought for themselves about who to pay attention to. Both indirectly accuse the other person of ineptitude.

Another sign a question is poor is if the question has a "right" answer. When someone jumps in and asks us "What were you thinking??" the answer they're usually looking for is some variant of "I wasn't thinking" followed by a quick apology for a perceived wrongdoing. Questions with correct answers aren't questions born of curiosity—they are questions borne of a desire to stroke our egos at the expense of someone else.

Bad questions usually lead to discussions and debates that go nowhere. My question to my son about jumping off a bridge if his friends did was a question with a right answer. In attempting to control his answer, I triggered psychological reactance, and my son responded purposefully with the "wrong" answer to assert his own autonomy. (Naturally, this did nothing to resolve our disagreement or bring us closer together.)

Of course, tone goes a long way too. If a friend is telling you a harrowing story and you ask "What were you thinking?", this may be a genuine desire to understand their thought process in a stressful scenario. Tone, timing, and relational context matter.

Good questions are powerful in a conversation. Good questions can shift the dynamics of a stuck conversation, opening up new possibilities for collaboration and compromise. Asking genuine, open-ended questions sparks curiosity, reduces psychological reactance, and builds bridges of understanding between opposing stakeholders in the energy debate.

I experienced this firsthand as part of a panel discussion at an energy conference. The panel included a utility executive, an environmental activist, an academic researcher, and myself. The moderator kicked off the discussion with a provocative question: "Is it possible to reconcile the need for climate action with the growing global demand for energy?"

The utility executive jumped in first, purporting that fossil fuels would remain essential for ensuring reliability and affordability, especially in developing countries. The activist countered that we couldn't afford to keep investing in dirty energy if we wanted to avert climate catastrophe. The academic offered a flurry of statistics on renewable energy costs and adoption rates. Within minutes, the conversation devolved into a heated back-and-forth, with each panelist trying to score points and prove the others wrong.

The moderator wasn't doing much moderating (which is frustratingly common with these type of events), so, out of sheer irritation, I decided to try a different approach. "Can I ask a question?" I interjected. "What do each of you see as the biggest opportunity for collaboration between the fossil fuel industry and the clean energy sector in the next five years?"

The panelists exchanged skeptical glances. The utility executive crossed his arms.

"Collaboration?" he scoffed. He probably thought I was nuts. "Our industries have fundamentally different goals and incentives."

The activist nodded in agreement. "Yeah, it's hard to see how we can find common ground when one side is actively destroying the planet."

These were grown adults operating in the professional world. I wanted to press my fingers to my forehead because they were all giving me a headache. Instead, I pressed on in the conversation.

"I hear your skepticism," I acknowledged. "But humor me for a moment. If you can't think of an opportunity to collaborate, what would make collaboration feel more possible?"

The utility executive, arms still crossed, shrugged. "Well," he started with a huff, "environmental groups could recognize the projects we're already doing, that would help. And if they were more open to solutions like carbon capture and nuclear power, and less reflexively anti-fossil fuel, that would be a good start."

The academic researcher spoke up. "For the record, I never said I was against nuclear power—Anyway, if fossil fuel companies were more transparent about their emissions and more willing to consider outside help when it comes to reducing those emissions, I would find that a show of solidarity."

The conversation that followed was far from easy—in fact, it was pretty painful. There were still plenty of disagreements and tense moments. But by the end of the panel, the tone of the conversation had noticeably shifted. With a few more good questions (mostly prompted by me, with the moderator catching on), it appeared both sides were becoming curious about solutions that would work for both of them, instead of arguing solely about the merits of their own ideas.

When an argument devolves into each person fixated on scoring figurative points, we go nowhere. While no one fundamentally changed their ideology because of this conversation (not that I expect this *ever*), everyone did seem to recognize dialogue was possible. And everyone left feeling a bit more hopeful about the possibility of change in the energy industry.

By reframing the conversation around potential opportunities and shared challenges rather than on long-standing disagreements, the tenor of the discussion shifted. The panelists started listening more intently to each other, building on one another's ideas rather than trying to tear them down. The audience, too, became more engaged, offering their own questions and suggestions.

This experience underscored for me the transformative power of good questions in shaping the trajectory of difficult conversations. Open-ended questions inviting curiosity and reflection are far more effective for building understanding and uncovering common ground than leading questions that put people on the defensive.

Some key principles I try to follow when asking questions in charged contexts include:

1. Focus on underlying common interests and values, not just stated positions. "What's most important to you in this situation?" instead of "Do you support this policy?"
2. Presume positive intent and capability. "How can we work together to address this challenge?" instead of "Why haven't you solved this yet?"
3. Acknowledge valid concerns. "What outcomes would we all like to see?" instead of "What's wrong with the other side's approach?"
4. Encourage creative problem-solving. "What possibilities might we be overlooking?" instead of "Which option do you prefer?"

When stakeholders on all sides of the energy debate learn to stay curious and ask these kinds of questions, it has a transformative effect on our conversations. The Radical Middle Mindset, plus good questions, break down the barriers of defensiveness (reactance) and opens up space for potential collaboration. When stakeholders can conduct level-headed conversations across interests, this enables nuanced, context-specific problem-solving. Instead of a focus on winning arguments, stakeholders can move the needle toward change.

While the Radical Middle Mindset and good questions are crucial to finding pragmatic solutions, it's also helpful to enter into conversations with a basic understanding of others stakeholders' incentives.

## UNDERSTANDING INCENTIVES

Psychologists have long recognized that behavior is heavily influenced by incentives. Motivation 101 asserts people and organizations tend to pursue actions that are rewarded and avoid those that are penalized. This is not earth-shattering theory, yet every time I break it down, people look at me like I am unveiling a cure for cancer. Asking good questions is awesome and effective—but there are also some fundamental incentives driving every type of stakeholder. Understanding these incentives can help give us a head start in coming to solutions.

In the complex world of energy policy, different stakeholders often appear to have conflicting incentives. Generally speaking, fossil fuel companies are guided by market forces and incentivized to optimize the balance between production and profits. Environmental groups are typically guided by a non-profit mentality and motivated to fundraise and exert their respective message, which in most cases, revolves around the reduction of emissions as quickly as possible. Policymakers are constantly pulled between short-term wins and long-term gains to satisfy their varied constituents. However, a Radical Middle approach recognizes that these surface-level conflicts often mask deeper alignments of interest. By getting curious about each party's underlying needs and constraints, it's possible to find hidden opportunities for mutual gains.

For example, while fossil fuel companies and environmental groups may clash over specific policies, they share a fundamental interest in the long-term viability and social legitimacy of the broader energy sector. No company wants to be rendered obsolete by disruptive innovation. And no environmental group truly benefits from demonizing the industry powering the global economy. Both sides have an incentive to work together toward increased energy efficiency. They probably will not admit it, but they probably share more in common than they think.

The key is to get away from positional bargaining and instead focus on interests, and there are promising examples of this interest-based approach yielding breakthroughs in the energy domain. For instance, the US Nuclear Energy Leadership Act brought together an unlikely coalition of unions, environmental groups, and industry players to support advanced nuclear research and commercialization.[2] By focusing on shared interests in job creation, emissions reduction, and US technological leadership, advocates were able to overcome long-standing divisions and achieve a major bipartisan victory.

As another example, Rocky Mountain Institute's Electricity Innovation Lab (e-Lab) initiative has convened utilities, technology providers, regulators and NGOs to collaboratively develop new utility business models. By creating a space for open, solutions-oriented dialogue, e-Lab has enabled parties with divergent positions to discover alignments around future energy system architecture and craft win-win solutions like performance-based regulation.

These examples point to the power of reframing energy challenges around goals that unite different constituencies. Identifying shared interests in core outcomes like affordability, reliability, security, air quality, and economic opportunity can break through the zero-sum mindset and unlock creative problem-solving. Stakeholders don't have to agree on everything, or even most things. They just need to filter out the tedious rhetoric, politics, and frankly, bullshit, and then find even a fraction of common ground to move forward on specific initiatives.

In the coming chapters, we'll explore dealing with the three main stakeholders in the energy industry: energy companies (which includes fossil fuel folk); environmentalists, and policymakers. We'll look at their unique motivations and how their individual industries perpetuate ideological thinking. Then, we'll learn how to better communicate with each group. There are particular (good) questions to ask and (bad) questions to avoid when in conversation with each stakeholder that will encourage them toward the Radical Middle.

# CHAPTER 6
# HOW TO BE CURIOUS ABOUT ENERGY STAKEHOLDERS

To work with energy stakeholders on impactful and pragmatic solutions, we need to first be curious about their operations and attain a high-level understanding of how energy actually works. While every individual in the energy sector will have slightly different incentives, there are some basic motivations shared by energy producers, power plant operators, and fossil fuel companies. We'll explore the factors on the mind of every stakeholder in the energy business, then consider how to approach these stakeholders with meaningful and thoughtful questions leading to a deeper understanding of their specific concerns.

Anyone trying to collaborate in the energy space—whether it's with a fossil fuel company, power plant operator, energy producer, or energy distributor—must understand energy stakeholders are firstly in *business,* and we do not get to magically change the laws of economics. In other words, it is critical to acknowledge and accept the fact that laws of economics dictate the actions of energy industry players. Therefore, they will be primarily concerned with what any business would be: (1) staying alive and (2) remaining profitable. Engineers, floor managers, and technicians within an energy company will have differing opinions on what their companies should or shouldn't be doing. But all major decision makers in any energy-related businesses will be primarily concerned with managing their finances to stay alive. If they are not, the company is likely to fold.

Although it might seem like energy companies are indispensable and therefore rolling in cash, not every energy company is swimming in dollars. In fact, it is often the opposite: most energy companies will struggle with thin margins. Sure, the likes of ExxonMobil, BP, Shell, and ConocoPhillips are well off. But most small to mid-sized producers (the Middle Guys) often survive on thin margins. The energy industry is over a hundred years old, and as such energy producers and power plant operators handle old equipment. Building and maintaining (not to mention upgrading) energy infrastructure requires massive amounts of capital, which most small and mid-size producers don't have lying around.

The first reason most energy stakeholders function on slim margins is that energy is a global commodity. Energy rates (and commodities like coal, oil, and LNG) are influenced by a larger market where prices are set by players across the world. The rhetoric continually accusing the industry of price manipulation is false. Let's say a mid-size US coal producer sells half their typical supply to domestic power plants and the other half they export. If the price of coal rises, great. They can certainly increase prices globally and generate profit from their exports. But domestically, they can't raise prices much because the power plants can only afford to pay within a set range. The coal producer might try to export more than half to take advantage of the rise in price, however, the power plant demands a certain amount of coal to literally keep the lights on. In this scenario, the price going up works well for the coal producer, at least for the exports.

However, if the global price of coal *falls*, the coal producer faces a separate set of issues. As the global price of coal decreases, so do the producer's profit margins on their exports. Their production costs don't change, but how much they can charge for half their product (the exported coal) goes down. Politicians and fossil fuel detractors tend to leave this part of the story out of their narrative. The coal producer might consider raising the price for their coal domestically to make up the difference, however, the power plant might not be able to afford the increase in price, and might consider finding another, cheaper supplier.

The quirks of the global market are just one factor creating uncertainty for an energy company—whether that's a fossil fuel company, a power plant, or a solar farm. And there are additional factors preventing a fuel producer or a power supplier from adapting to market changes the same way another business might.

While energy companies can pass some costs onto the customers, energy prices in many regions are regulated energy prices so energy companies can only raise prices within a fixed range. This limits a company's ability to pass on costs to customers. Energy companies are unique in the sense that they can't operate like the consumer businesses most people think of. If demand for Nike sneakers suddenly jumps, Nike (and any seller of Nike sneakers) can increase prices as much or as little as they want. There are no restrictions on how much they can sell sneakers for. But energy can't charge whatever they want when demand goes up. In fact, they could face legal trouble if they hike rates over a certain threshold because energy prices are monitored and regulated by government agencies.

Several government agencies oversee energy prices for ratepayers. The Federal Energy Regulatory Commission (FERC) oversees and regulates rates to ensure they are fair on a national level. Additionally, every state in the US has a Public Utility Commission (PUC) or a Public Service Commission responsible for regulating utility rates to ensure rate prices are just and reasonable. Consumer goods are overseen by the Federal Trade

Commission (FTC), but again for many businesses outside of energy, they can charge whatever the market (and customers) will withstand.

Because everyone needs access to reliable energy in their daily life, it is wise to have some restrictions on what energy companies can charge. Without guardrails, energy companies might increase rates to whatever number they want overnight, jeopardizing the lives of millions. (If this sounds drastic, consider how deadly intense heat waves can be, or the risks of hypothermia. Some routine medications need to be refrigerated, and hospitals have to pay for power just like everyone else.) I'm not here to pass judgment on these specific regulations—however, the fact that the energy sector is already highly regulated impacts the ability of energy companies to be dynamic in their business plans.

With tight pricing restrictions and variable global markets, energy firms must be ruthlessly efficient with their budgets and capital deployment strategies. Most do not have the excess capital to consider upgrading their infrastructure or technology let alone innovate or be environmental stewards. Investments in emissions reduction technologies, carbon capture, or offsetting programs are luxuries mostly only affordable to the biggest players (e.g., BP, Shell, ConocoPhillips, and other massive energy conglomerates). It's not that smaller participants don't care—most don't have the balance sheet to facilitate such innovation.

Unlike consumer-facing businesses that can grow virally, energy is an infrastructure-based industry where change happens gradually. And change happens slow, mostly because of all the regulatory bureaucracy riddling the industry. Let's remember that much of the U.S. grid was built in the 1960s. The last time the United States focused on building out the grid was when music fans organized Woodstock. Our aging infrastructure is struggling to meet our modern electricity needs today. According to the U.S. Department of Energy, 70 percent of transmission lines are over twenty-five years old and approaching the end of their typical lifecycle[1]. We need to address the immediate consequences on the horizon, namely power outages, susceptibility to cyber-attacks, and existing faulty grid infrastructure, before we set lofty energy goals that, lets face it, often serve a political agenda

Energy companies need to shore up massive amounts of capital to create the change policymakers and environmentalists demand. Policies divesting funds from fossil fuel and energy companies often backfire—the small and mid-sized players need capital infusions to upgrade and innovate if we expect them to reduce emissions. Divestment from fossil fuels becomes counterproductive to creating change in the energy industry.

To partner with energy companies, environmentalists and policymakers must understand the primary motivation for these companies is to stay alive and increase profit margins. The Radical Middle approach recognizes that energy companies are not monolithic villains, but collections of people responding rationally to the incentives in front of them. Policymakers and environmentalists can aim to frame collaborations as giving energy companies what they need and want: cushion to create long-term growth. Writing energy companies off as the bad guys discounts the enormous pressure the bulk of the energy sector (i.e., small to mid-sized producers) faces just to survive from one quarter to the next.

Policymakers and environmentalists tend to assume what energy companies' motives and incentives are. Often these groups think energy companies have swimming pools full of money the likes of which Scrooge McDuck could only dream of. But many—most even—are struggling to get by. It's easy to slip into accusatory language or leading questions that put energy companies on the defensive. If we want to make progress, we must be intentional about how we frame our inquiries. We need to ask better, and more relevant, questions of energy stakeholders.

## Q&A WITH ENERGY COMPANIES

As we discussed in the previous chapter, curiosity is not the only crucial ingredient for meeting in the Radical Middle. We often inadvertently trigger psychological reactance in others (and feel it ourselves), and it is a major barrier to productive dialogue. When people feel their freedom or autonomy is being threatened, they often become defensive, dismissive, even hostile. In energy debates, it's all too easy for a benign inquiry to backfire by triggering reactance, so how we phrase questions matters.

One of the most effective ways to engage energy companies in constructive dialogue is by asking open-ended questions demonstrating a desire to more deeply understand perspectives, challenges, and motivations. By inquiring about the complex realities energy companies navigate, we can begin to identify shared goals and mutually beneficial solutions.

Of course, not all questions are equally effective at fostering productive exchanges. Asking an oil executive an open-ended question like "Could you tell me how you sleep at night knowing your business is destroying the planet?" is more likely to provoke a sarcastic retort than a thoughtful reflection. These types of loaded, judgmental questions only reinforce an adversarial dynamic.

Instead, try asking questions inviting energy leaders to share their perspectives, insights, and experiences on the challenges and opportunities they see. Here are some examples, in no particular order:

**What do you see as the most promising technologies or strategies for increasing energy efficiency while meeting growing global energy demand? What market conditions would help accelerate deployment of energy efficient technologies and infrastructure?**

Energy efficiency is good for energy companies, ratepayers, and environmentalists, so framing questions around moving toward energy efficiency is typically a good way to start a pragmatic conversation. These questions assume energy companies all want to do better (because it's good for business), and quickly move on to focus the conversation on pulling out the unique perspective of this energy company on *how* to do better. In other words, these questions ask the energy companies to focus their answers on actionable measures to progress toward doing better business, which includes mitigating negative externalities (like emissions).

**Where do you see the biggest opportunities for updates in energy?**

This question allows the energy company stakeholder to take the lead and explain to you where their thinking is when it comes to necessary change in their industry. Instead of barreling in with your opinions on the biggest opportunity and attempting to force them to take your perspective, you can first see how they think about progress in energy, and then work toward finding common ground.

**How do you engage with local communities and other stakeholders in the areas where you operate? What approaches have you found most effective for building trust with them and addressing their concerns?**

This question recognizes how energy development can significantly impact local environments and communities. It invites companies to share examples of how they are connected to and navigate in their communities. If you're speaking with a representative from a small to mid-sized energy company, questions like these are great for better understanding their role in the local economy. For larger energy companies, you may need to ask these questions of the directors and executives in direct contact with the local operators.

**What do you see as the appropriate role of federal and local government as you try to update your infrastructure and maintain operations? What sort of incentives would be useful to you?**

These questions open the door to conversation on energy companies' view on the balance between government intervention and market forces. Most will probably respond with a snarky statement of how the government should butt out of their business. However, the government is already involved in the energy sector, and it will be difficult to disentangle their influence. These questions acknowledge the role of government and ask energy companies to instead focus on what would be helpful to them by way of government incentives.

**What do you see as the biggest challenge(s) for you in keeping up with the impending demand for energy as data centers and tech demand more energy?**

Not every energy company might be thinking about the increasing demand for energy—but they all should be. This question helps you understand if they are thinking about it at all and how exactly they see their operations being impacted.

**How is your company investing in workforce development and training to prepare for the energy demands of the future?**

Asking about workforce issues signals an understanding that the energy transition will have major implications for jobs and livelihoods. It provides an opening to discuss how companies are managing the everyday nuts and bolts of running a business.

**How are you leveraging digital technologies and data analytics to optimize your operations and improve efficiency? What sorts of analytics would be most helpful to you?**

This question acknowledges the critical role of digitalization and innovation in modernizing the energy system. It invites companies to share examples of how they are harnessing these tools to drive improvements and what sorts of resources would be most beneficial. Answers to these questions can lead to finding solutions and opportunities for collaboration.

**What lessons have you learned from past efforts to develop and deploy new energy technologies? How have**

**those experiences informed your current strategy and investments?**

This question recognizes that the history of energy innovation is marked by both successes and failures. It invites companies to reflect on what has worked well and what hasn't, and to share insights that could help accelerate progress going forward. It also offers energy companies the opportunity to show they are not monolithic and indeed do take action to improve their operations—it's just often behind the scenes.

**How does your company think about environmental stewardship when the budget allows for it?**

Asking about environmental stewardship is generally risky because it's such a hot button issue for so many people. However, this question is phrased in a more open manner that does not assume a company must consider environmental stewardship. This question gives the energy stakeholder a graceful way out, offers them the chance to express their views on the idea of environmental stewardship, without requiring them to fall on one side of the climate debate or the other.

By posing questions like these, we can gain valuable insights into the complex factors shaping energy companies' decision making. We can also better understand the risks for all companies on a relative scale. Whether it's a one-on-one conversation, a panel discussion, a question in a public forum, or a business conversation over dinner, these questions avoid psychological reactance, and allow the asker to better understand how energy companies operate and think about innovation.

We can all identify areas of common ground and explore pragmatic solutions that work for more than one involved party. Plus, leading with curiosity is a more effective way to drive change than leading with condemnation. If energy companies, policymakers, and environmentalists are serious about creating change, they're going to have to ask better questions before they can collaborate.

## COLLABORATING WITH THE ENERGY INDUSTRY

Too often, environmentalists and policymakers make the mistake of treating energy companies as outdated entities that can simply be ordered to change. Both groups issue blanket demands or propose one-size-fits-all solutions without regard for the complex economic, technical, and political factors shaping corporate decision-making. When their initiatives meet resistance or fall short, they're surprised and frustrated.

The truth is, even the most forward-thinking energy executives are bound by the constraints of their balance sheets, their fiduciary duties to shareholders, and the competitive dynamics of the market. They can't simply flip a switch and become a different kind of company overnight, no matter how much we might wish they would. Change in this industry happens incrementally, through strategic investments, reasonable policy reforms, and shifting global market dynamics.

To collaborate effectively with energy companies, we need to put ourselves in their shoes and look for win-win opportunities. We need to understand that every company, first and foremost, will prioritize economics (which includes mitigating negative externalities such as polluting the local environment), and will be focused on protecting its capital reserves and ensuring its own survival. They can only start investing in innovation, environmental stewardship, and social responsibility from a foundation of financial viability.

So, what does this alignment of incentives look like in practice?

Consider a hypothetical scenario where environmentalists want to work with a major utility to accelerate the retirement of coal-fired power plants in their service territory. A knee-jerk approach might be to launch a public shaming campaign or lobby for regulations mandating plant closures by a certain date. But this adversarial strategy is likely to trigger fierce resistance rather than cooperation. Alternatively, the environmentalists might demand a change over to renewables immediately, maybe even flirting with a rival power provider eyeing the region for development.

An alternative Radical Middle approach would be to first seek to understand the utility's perspective. What are the economic barriers to early retirement of these assets? How might shutdowns impact reliability, affordability, and the local workforce? Are there legacy obligations or contractual constraints that need to be addressed?

Armed with this context, environmentalists could then work with the utility to design a phased transition plan balancing climate goals with customer and shareholder interests. This might include strategies like:

- Advocating for favorable tax treatment or access to grants and low-cost capital for increases in energy efficiency.
- Exploring opportunities to retrofit coal plants to run on cleaner fuels like natural gas.
- Developing targeted job training and economic diversification programs for impacted workers and communities.

The key is to approach the situation as a shared challenge to be solved collaboratively. By aligning with an energy company's main priority—economic viability—environmentalists can take one big step forward.

This principle of enlightened, pragmatic engagement extends beyond just coal plant retirements. On any number of issues—from deploying renewable energy and electric vehicle infrastructure to improving energy efficiency and grid resilience

—there are opportunities to find creative compromises advancing multiple goals simultaneously.

Another illustrative example is the growing trend of major oil and gas companies investing in renewable energy tech. At first glance, this might seem like greenwashing—an attempt to burnish their public image without fundamentally changing their business models. And indeed, some of these efforts may be more PR-driven than transformational. However, seen in another way, these investments reflect a rational response to shifting market conditions and an attempt to diversify risk. If the big players don't stay ahead of innovation, they jeopardize the longevity of their company, leaving themselves excluded from the big gains disruptive tech could have on the energy industry.

By getting curious about the strategic motivations behind these moves, clean energy advocates can explore ways to leverage fossil fuel companies' vast expertise, capital, and infrastructure for energy efficiency innovations. (The U.S. energy space is also home to some of the best engineering talent on the planet, which could be better utilized.) This could take the form of joint ventures to develop utility-scale projects, research and development (R&D) partnerships to advance energy storage and grid integration technologies, or pilot programs to test new energy tech in real-world settings.

Overall, conversations with energy companies should revolve around finding synergies and accelerating the most promising solutions. This means advancing pragmatic solutions (often full of compromises) that are viable in the real world. Ideological purity might make us look good to our in-group, but it isn't useful in creating *action,* and it runs counter to figuring out solutions.

The Radical Middle approach involves channeling capabilities in a more sustainable direction by finding areas of aligned interests. It also involves cultivating an ecosystem of cross-sector partnerships and unorthodox alliances in service of shared goals. Finally, it involves being pragmatic and adaptive in the face of complex, evolving challenges.

But we need to better understand not only energy companies but also the two additional parties in the energy debate: environmentalists and policymakers. We'll move next to environmentalists.

# CHAPTER 7
# HOW TO BE CURIOUS ABOUT THE ENVIRONMENTALISTS

Moving forward, to effectively collaborate with environmentalists in this new world, it's essential to first understand the motivations shaping their behavior. Just as energy companies are constrained by economic realities and fiduciary duties, so too do environmental groups operate within a set of institutional imperatives and stakeholder demands. We'll first take a closer look at environmental NGOs, then focus more closely at the pressures on those in the climate research (academic) field and renewable tech space.

At the most basic level, environmental NGOs are mission-driven organizations dedicated to protecting natural resources and public health. Their purpose, as many of them see it, is to advocate for cleaner air, water, and landscapes, and to combat the polluting practices threatening these shared goods. This core identity exerts a powerful pull on organizational strategy and messaging.

For many groups, an adversarial stance toward fossil fuel interests is seen as central to their advocacy role. Opposition to coal, oil, and gas development is a point of pride for most people in environmental NGOs. Activists who appear too compromising with fossil fuels in their interactions risk being labeled as greenwashers. Because of this, many maintain hard-line stances against fossil fuels.

This pressure to maintain an uncompromising stance is reinforced by the expectations of environmentalists' constituents and, perhaps most importantly, funders. Most green groups rely on donations and foundation grants to sustain their work. This financial model incentivizes bold, uncompromising positions that energize the base, attract media attention, and drive successful fundraising efforts. Many of the major environmental donors, whether large foundations or wealthy individuals, have grown impatient with incremental approaches and are pushing for more disruptive, systemic change strategies.

Because of the financial pressure from the biggest funders, environmentalist NGOs have moved away from grassroots efforts focused on affecting real change in local communities. Instead, the NGOs spend most of their time and resources pandering to the biggest donors so they can survive from one quarter to the next. They face pressure to demonstrate big gains or risk losing the confidence and support of key benefactors.

With countless groups vying for limited funding and mind-share, the pressure to differentiate and justify one's existence is intense. Staking out the most radical, uncompromising positions is one way to cut through the noise and demonstrate value to donors and allies. This can be seen in the wave of public disobedience displayed by an array of environmental activists. Fervor over impending climate doom is typically good for boosting donations and recruiting volunteers.

Of course, it is near impossible for NGOs to create sweeping radical change all at once on a shoe-string budget. They might be able to stoke support and media attention, but fewer NGOs are good at creating actual change—because actual change is usually slow and incremental. It's also difficult to tell people there's too much $CO_2$ in the air, when they are breathing just fine. Selling climate change is abstract, and many NGOs have forgotten their grassroots roots. The focus of environmental

NGOs has shifted to focusing on metrics catering to quasi-elites like quantifying carbon and the speed at which glaciers melt. Forgotten are the more tangible issues affecting people on all rungs of the socioeconomic ladder.

Before, there were environmentalist groups with lofty abstract goals, there were grassroots environmentalist movements creating measurable change. Smog cleared, rivers became clean enough to swim in, and forests came back to life thanks to concerted efforts by environmentalist advocacy groups. Grassroots activities were targeted toward specific tangible goals—e.g. reduce the amount of toxic waste in a river so people can safely swim and fish in it. Goals were well defined, and success was tangible and easily measured.

Today, the biggest environmentalist groups typically focus on emissions reduction targets. These are hard to measure globally and for most people are abstract. However, the advantage with this goal is environmentalists can play into the us-vs-fossil-fuels narrative that rallies constituents.

Of course, not all environmental groups are solely focused on fossil fuel opposition. Many have developed reasonable policy agendas around energy efficiency, renewable energy deployment, consumer use reduction, and other critical transition strategies. Some have even forged pragmatic partnerships with corporate actors to accelerate adoption of energy efficient tech and practices.

Yet, even for these more solutions-oriented organizations, the need to bring in funding can still create pressure to support extreme-ends policy measures and generate pressure campaigns rather than engage with the energy industry. In today's media, collaborative solutions are boring. Celebrating partial victories or incremental progress is difficult in a media environment rewarding conflict and overnight successes. And in the aftermath of decades of fossil fuel industry intransigence and misinformation on climate change, the burden of proof is high for any environmental group seen as collaborating with the "enemy."

(And indeed, there is plenty of environmentalist in-fighting, which ironically, undermines the very efforts they claim to be promoting—more on this later.)

Environmentalists in the academic world face similar funding pressures as NGOs. Many research labs rely on grant funding or donations to continue work. This can result in a bind for academics. On one hand, funders and grants demand they find evidence to back an extreme end of the climate debate. Yet the very nature of research is about better understanding nuance and specifics. Within the environmental studies field, researchers are pushed to take hardline stances. Funding pressure can push academic research to support extreme ideologies, which in turn perpetuates us-versus-them narratives.

Briefly, renewable tech companies (particularly startups) face similar pressures when it comes to funding and support. Often, they face immense pressure from funders to make the most disruptive tech possible (which would secure the most financial gains for investors). In crowdsourcing scenarios too, renewable tech companies usually cater to an extreme end of the spectrum to curry more donations from supporters. In both scenarios, the "disruptive" aspect of potential renewable tech positions these companies as enemies of the current energy industry. Unfortunately, because of the pressure to create the most disruptive "green" tech possible, many companies miss out on opportunities to learn from and collaborate with current energy industry players.

This is the terrain environmentalists must navigate as they chart a course through the energy sector. They may wish to explore opportunities for collaboration, but pressure from donors and other environmentalists, even some policymakers, forces them toward extreme stances. It's renewable tech or nothing, so the narrative goes. This makes any sort of progress difficult as they demand rapid change from an energy system deeply entangled with incumbent industry players and resistant to *quick* change.

Armed with an understanding of the basic pressures faced by environmentalists, policymakers and energy companies can begin to dig deeper with their questions. Collaborating with environmentalists can be difficult, and the most extreme-minded environmentalists are more focused on proving their narrative correct under any circumstance. Conversations with environmentalists will first require you to gauge how far away from the Radical Middle they are, then work on questions that would encourage them toward the middle. This approach can also help environmentalists when in conflict *with each other* over initiatives and policies. Let's take a look at what those questions might be.

## Q&A WITH ENVIRONMENTALISTS

As we've seen, environmental groups are driven by incentives and constraints that can make engaging in good-faith dialogue challenging, if not near impossible. The combative posture and uncompromising rhetoric many advocates feel compelled to adopt are often at odds with the kind of nuanced, solutions-oriented approach needed to drive real progress.

The environmentalist camp will have a range of moderates to extremes, but they all believe climate change is happening. There is no point arguing with them about whether or not climate change is "real." There is evidence to support almost every variation on the severity or triviality of climate change. Arguing the scale and depth of a climate crisis is a waste of time. At all costs, resist falling into the trap of ideological debates. If the environmentalist keeps bringing up the validity of climate change, avoid a discussion about it, and redirect attention toward solutions and finding common ground.

Here are some examples of constructive questions that can help foster a more productive dialogue with environmental leaders and advocates:

**How do you prioritize which environmental issues to**

**focus on (e.g. air pollution, water contamination, land conservation, climate change)? How do you determine where you can have the greatest impact given limited resources?**

These questions invite environmentalists to share their decision-making frameworks and strategic rationale. By drilling down into specific issue areas, it grounds the conversation in concrete challenges rather than abstract ideologies and helps the question asker understand the environmentalists' priorities. In particular the second question can reveal what data and analytics sources environmentalists attend to most.

**What partnerships will be needed to drive progress in energy use among vital economic sectors like agricultural, industry, construction, and real estate?**

This question first requests environmentalists think in areas that are energy heavy and vital to the economy, yet not often discussed. With a focus on potential partnerships, there is an assumed need for collaboration which helps avoid answers that might only focus on what other groups (not environmentalists themselves) should be doing individually.

**How do you craft initiatives and campaigns to be mutually beneficial between stakeholders in the energy industry?**

By framing the goal as mutual benefit, this question discourages rumination on what's not working (which is likely to trigger anger and blame). Instead, answers should revolve around discussing *actions*. If an environmentalist doesn't have an answer, this question sheds light on how they have neglected to attempt collaboration. And it pushes them to grapple with the needs and constraints of different constituencies.

**What lessons have you learned from past collaborations with the energy industry and what would you do differently in future engagements to build trust and find common ground?**

Asking advocates to reflect on their history of working with industry demonstrates an interest in learning from past successes and failures. It also acknowledges building productive relationships takes time and effort. A focus on trust and common ground, it sets a forward-looking tone and invites advocates to suggest strategies for improving communication and alignment. It also creates space for them to air grievances with tangible examples to back up any accusations.

**How do you think about balancing urgent environmental needs with keeping costs low for ratepayers? What strategies do you use to ensure communities both near and abroad are centered in your advocacy?**

This question gets at the heart of the tension between environmental and economic imperatives. It acknowledges the legitimate concerns around energy affordability and access, especially for disadvantaged groups, which environmental NGOs are often concerned about given most climate activists' adjacent focus on climate ethics. By asking advocates to share their approach to managing these trade-offs, it invites a nuanced discussion about maintaining standard of living while innovating and upgrading the energy sector.

**How do you navigate tensions between short-term and long-term priorities in your advocacy? How do you maintain momentum and stakeholder support for initiatives that may take years or decades to bear fruit?**

Highlighting the challenge of time horizons normalizes the struggle many advocates face in their work. Asking how they sustain momentum over the long haul elicits valuable insights about their goals regarding communication, coalition-building, and adaptive strategy. It also assumes change will take time, a fact the most hardline environmentalists often forget.

**How does your organization think about the global dimensions of the clean energy transition, particularly the need to expand access to energy in developing countries? What responsibility do you feel nations, corporations, and NGOs have to support this process and how can it be done pragmatically?**

This question pushes environmentalists, who often are

focused on the Western world's energy consumption, to think globally about energy and its environmental and economic impacts. Energy policy doesn't only affect domestic goings on, it affects nations worldwide, and poor policy has lasting ripple effects and a slew of unintended geopolitical and socioeconomic consequences. A growing energy demand isn't only a US issue —it's a global one.

Developing nations like China, India, Venezuela, Brazil, and the Philippines are just getting going with energy access for all citizens. What's available (and cheap)? Fossil fuels. Additionally, "green" tech deployment relies on a fairly sophisticated infrastructure, which many developing countries don't have. This question requires environmentalists to examine the practical problems faced by nations trying to expand or maintain access to energy for standard of living and come up with new solutions.

**What's important to you when communicating the impacts of your work to funders and the public?**

Answers to this question reveal what environmentalists want to show off when they are touting their achievements. It also helps us understand what sort of goals they may be trying to reach to please their constituents. Understanding environmentalists' goals can help us better understand what benefits they might look for when making compromises.

**What do you see as the most promising opportunities to accelerate change in the energy sector? What policies, technologies, or market trends give you hope?**

This question shifts the conversation into a positive, solutions-oriented frame. It invites advocates to share their vision for the future and highlight what they see as examples of progress. This helps us better understand what they see as success and find opportunities for collaboration. It also creates space for advocates to showcase their expertise and propose innovative strategies for driving systemic change.

**How do you define success for your organization? What metrics do you use to track your impact and effectiveness?**

Asking advocates to articulate their definition—and particularly their metrics—of success pushes them to get specific about their goals and priorities. It invites them to share the indicators they use to assess their performance and make strategic decisions. By grounding the conversation in tangible metrics, it moves beyond lofty rhetoric to a more concrete discussion of what it will take to achieve meaningful results. It also provides insight into how environmentalists think about accountability and continuous improvement.

These questions are designed to invite contemplation and disclosure rather than defensive posturing. They create space for environmentalists to articulate their values, share their perspectives, and confront the real-world trade-offs and dilemmas they and others face creating change in the energy industry. And, of course, these questions allow us to better understand environmentalists' viewpoints so that we might better prepare for collective action.

It's not a magic bullet and it won't work for everyone, but following these suggestions will give you the best chance to engage in productive and meaningful discussions with environmentalists.

It's also worth noting not every environmentalist will be receptive to this kind of open-ended dialogue. (More on what to do when this happens in Chapter 9.)

With the right questions, instead of working against environmentalists, we can harness the energy and education of their movement to drive progress.

## COLLABORATING WITH ENVIRONMENTALISTS

Engaging in dialogue to understand environmentalists' nuanced perspectives is an essential first step toward collaboration. Despite what environmentalists may think, not every fossil fuel company is full of oil-loving die-hards. Many people in the energy sector recognize and would be happy to help protect the environment—we all just have different perspectives on what to prioritize.

Despite such varied perspectives, it is possible for seemingly opposing sides to collaborate. A few examples include Clean Energy Buyers Alliance (CEBA), the National Carbon Capture Center (NCCC) in Alabama, and the Low Carbon Fuel Standard (LCFS) out of California. While it would be a stretch to say any of these are perfect, they are all proof that it is possible to collaborate if we are all willing to compromise on some of our ideological viewpoints in favor of action.

Clean Energy Buyers Alliance (CEBA) is a consortium of environmental NGOs, energy providers, and major corporate customers working together to accelerate change in the energy sector. What has made CEBA effective so far is its focus on creating win-win outcomes for all involved. Since its founding in 2014, the alliance has grown to include more than 300 members and facilitated over 50 gigawatts of clean energy deals.

Although CEBA's website and press releases do not reveal its particular successful actions, the existence of the consortium is the first step to collaboration. Fortunately, other institutions have better demonstrated the possibility of action. The National Carbon Capture Center (NCCC) in Alabama offers a promising model.

Established in 2009 with support from the U.S. Department of Energy, the NCCC is noteworthy for the diversity of its sponsor group. Major fossil fuel companies like ExxonMobil and Southern Company, as well as environmental NGOs like the Clean Air Task Force and the Environmental Defense Fund, have worked with the NCCC on carbon capture, utilization, and storage (CCUS, or simply carbon capture).

As an institution, the NCCC helps build trust and buy-in from key stakeholders on all sides. It does so by being a neutral platform for evaluating the technical and commercial viability of carbon capture solutions. The mission works because it aims to determine how to turn captured carbon, utilization, and storage into a money-making operation for fossil fuel companies. This goal also works for environmentalists because it reduces emissions. In fact, the NCCC spearheaded the formation of the National Carbon Capture Leadership Council, a coalition of industry, labor, and environmental groups advocating for federal policies to accelerate carbon capture technology and bolster public support for it.

Beyond project-based collaborations, there is also great potential for environmentalists and energy companies to work together on policy design and advocacy. In California, for example, the Low Carbon Fuel Standard (LCFS) has generated millions of dollars of economic activity, from fuel innovation to carbon credits to taxes.[1] The LCFS was developed through an extensive multi-stakeholder process that brought together environmental groups, fuel producers, vehicle manufacturers, and other key players to hash out the details of the policy.

While not perfect, but certainly trending in the right direction, the LCFS has been hailed as a model for how to craft durable, performance-based regulations aligning the interests of industry and advocates. The LCFS has unleashed a wave of innovation and competition in the clean fuels sector. It largely has done so by letting the free market determine the most cost-effective routes to success.

Other states and regions are now looking to replicate the LCFS success, with initiatives like the Pacific Coast Collaborative and the Transportation and Climate Initiative working to develop similar fuel standards across the West Coast and the

Northeast. These efforts have brought together an impressive roster of industry and NGO partners, from Shell and Chevron to the Union of Concerned Scientists and the Coalition for Renewable Natural Gas.

In addition to bringing together cross-sector stakeholders, being curious about environmentalists could serve environmentalists themselves. The environmental movement's hardline stance has led to infighting over methods and campaigns seen as too lenient. Many environmentalists who attempt to compromise or collaborate with energy partners are lambasted in the media and by their own supporters. Collaborators are treated more like Benedict Arnold rather than Nelson Mandela. Accusations of greenwashing and selling out to polluting corporations can threaten an NGO's funding and support.

Take the carbon markets as an example. Carbon markets have become a market for companies or organizations to offset their emissions profile with carbon credits. Instead of reducing their emissions, a company can pay to negate their carbon footprint. NGOs with environmental concerns have popped up to help regulate the market.

Yet, these carbon market NGOs are hotly debated. Some critics say offsetting only encourages companies to stay stuck in their ways, with the most vehement detractors will accuse carbon market NGOs of aiding corporate greenwashing. Supporters of carbon capture, utilization and storage (CCUS), however, suggest companies doing *something* is better than nothing.

It's not just energy companies, policymakers, and environmentalists who struggle to communicate with each other—each group has difficulty communicating among their own groups. With groups being pulled toward extreme ends of the ideological spectrum, communication with one's in-group can become contentious. It might seem like environmentalists all share the same viewpoint. However, each environmental NGO, academic, and activist will have slightly different perspectives on the state of climate and what solutions are best.

This is why asking good questions and approaching with curiosity instead of judgment is so important: the nuances under the environmentalist umbrella allow room for collaboration. But if we judge others as "pro-climate" *or* "anti-climate" we immediately limit the possibilities for productive conversation and collaboration. With the Radical Middle approach, collaboration across groups—and perhaps more importantly within our own groups to make measurable progress—becomes possible.

Now that we know how to better approach the environmentalists and energy industry players, one group remains, which is perhaps the most difficult to work with of all: the policymakers.

# CHAPTER 8
# HOW TO BE CURIOUS ABOUT POLICYMAKERS

Policymakers are perhaps the easiest group to judge—it's easy to think we know how government works. Yet, often when we assess policy decisions, we overlook and are unaware of the multitude of factors at play. Policymakers juggle constituent demands, local impacts, as well as global impacts.

Elected officials are beholden to the interests and demands of their constituents, at least in theory. The United States is a representative democracy (i.e. a republic), and policymakers always consider how their actions will be perceived by the voters who put them in office…and have the power to remove them. But that's not the only group policymakers must consider.

Policymakers are tasked with the delicate balancing act of appeasing several competing constituent groups and interests, including voters, campaign donors, their fellow party members, their party's platform, and local, national, and global impacts. While large groups of voters elect policymakers, many policymakers make promises according to what only a few, most generous campaign donors want. After all, campaign donations make it possible for the policymaker to reach the voters who put them in office.

There is also the perpetual pressure on policymakers to stay within party lines. Policymakers are part of greater political parties and may have to contend with differences within their own party. Whether campaigning or pushing forward an initiative, policymakers must consider the stance of their greater political party and balance macro-perspectives with local perspectives. A Republican in Florida may have a different platform than a Republican in Wisconsin; a Democrat in California may have a different platform than a Democrat in Kentucky.

There is no way around being a part of a political party. Historically, to remain successful, policymakers run and operate with the consistent backing of a political party. This allows voters to recognize some basic stances of the policymaker, regardless of the policymaker's personal opinions. Policymakers can more quickly garner support by being a part of a political party than by running independently. New policymakers might receive support or endorsements from retiring policymakers, boosting the chances the new policymaker might be elected. Additionally, many vote according to their party affiliations without looking too closely at a candidate's specific platform. Being a part of a political party is typically a boon for a policymaker.

Policymakers are often compelled to run on platforms capturing the most attention from the public and media. This usually means taking a hardline stance on issues and making audacious promises to fix "broken systems." The media, driven by a culture valuing conflict over consensus, often overlooks policymakers who make reasonable promises. While more pronounced in state and federal policymakers, this phenomenon also affects local policymakers to some extent.

Policymakers representing states or districts with significant fossil fuel industries face immense pressure to protect workers' livelihoods in these sectors. Policies seen as threatening to coal miners, oil rig operators, or pipeline builders can create resentment among voters, even if the policies offer long-term benefits in terms of energy efficiency and decreased energy prices. On the other hand, in more green-minded districts, there is pressure for policymakers to incentivize renewable infrastructure, which could lead to increased energy prices locally. Rarely does policy please everyone.

Even if policymakers find the perfect balance in regulatory action for their voter base and party, they still have to contend with policy's global effects. In today's interconnected world, energy policy decisions made in one country can have ripple effects across the globe. The United States, as the world's largest economy and second-largest emitter of greenhouse gases, has an outsized influence on international energy markets and geopolitical dynamics.

For example, when the U.S. Congress lifted the 40-year ban on crude oil exports in 2015, it sent shockwaves through the global oil market. Suddenly, American producers could compete directly with OPEC members for market share in Asia and Europe. U.S. crude oil exports increased from less than 500,000 barrels per day in 2015 to almost 3 million barrels per day by 2019.[1] This shift in the balance of power had far-reaching implications for everything from the price of gasoline to the short-term stability of petrostates like Russia and Venezuela. The increased supply of U.S. oil put downward pressure on global oil prices, impacting the revenues and economic stability of countries heavily dependent on oil exports.[2]

Similarly, U.S. policies around energy trade can significantly impact global economic and security conditions. For instance, restrictions on Iranian oil exports in 2018 squeezed Tehran's finances and raised tensions in the Middle East.[3] Halting liquid natural gas exports to the United Kingdom and the European Union could result in a greater reliance on energy sources from Russia, a political power many European countries are trying to maintain distance from, given Russia's unprovoked military attacks on Ukraine. The same holds for coal. In 2023, 70 percent of Consol Energy's (US producer and exporter of coal and a leading energy company) revenue came from selling its product abroad.

Policymakers must continually weigh complex geopolitical and economic considerations alongside domestic pressures from industry groups, labor unions, consumer advocates, and environmental organizations. They must grapple with the fact that the decisions they make about energy policy can profoundly impact the lives and livelihoods of people halfway around the world and the long-term habitability of our planet.

Making matters even more complicated, energy policy is often highly technical and multifaceted, requiring lawmakers to navigate a dizzying array of acronyms, jargon, science, and competing expert opinions. From arcane details of electricity market design to the minutiae of tax credit eligibility rules, policymakers often must make high-stakes decisions based on limited information and understanding. And because a policymaker's time is limited in office, they usually lack the time and bandwidth to acquire deep knowledge about the power grid.

In this context, it's no wonder energy policy debates can become so polarized and gridlocked. Finding common ground and forging durable compromises with competing interests and priorities can seem impossible. Policymakers who try to stake out a middle ground or engage in good-faith negotiations with the other side risk being labeled as traitors or sellouts by their party. And being ousted by one's party is a significant concern for a policymaker.

Converging in the Radical Middle is probably the most difficult for policymakers as they juggle competing constituent needs and complex economic and geopolitical variables. However, particularly for policymakers in office, whether at the beginning or end of their term, meeting in the Radical Middle is crucial to progressing on objectives. Policymakers who possess proof of action can better increase voters', donors', and constituents' confidence in reaching future goals.

Policymakers are challenging to work with. They are often

stretched thin for time and resources, and they do not trend toward objectivity. The amenable ones may make promises they can't follow through on, and the worst ones rarely listen to anyone but themselves.

However, there is hope. Speaking to a policymaker using the Radical Middle Mindset can help lower their defenses. Through good questions, we can demonstrate to policymakers we aren't out to "catch" them in a compromising position. Instead, we seek to understand their position so we can better achieve our needs alongside and with the aid of the policymakers.

## Q&A WITH POLICYMAKERS

One of the most effective ways to reach policymakers is by asking thoughtful, open-ended questions inviting policymakers to share their perspectives, experiences, values, and even, in some cases, their specific agenda. By understanding where they're coming from and what motivates their decisions, we can identify areas of alignment and opportunities for collaboration.

Of course, not all questions are created equal. Just as accusatory questions don't work to communicate well with energy stakeholders and environmentalists, nor do accusatory questions work with policymakers. Asking a policymaker something like "Why do you hate coal miners?" or "How much money did Big Oil pay you to vote against this bill?" is unlikely to yield a productive conversation. Accusatory questions trigger psychological reactance, putting policymakers on the defensive and shutting down constructive dialogue.

Instead, good questions to ask policymakers will demonstrate a genuine interest in learning a policymaker's priorities and how they conceptualize their role in forming regulation. You can use your questions to prove you understand their job is multifaceted and requires much deliberation regarding economics, politics, and civic duty.

Here are some examples of questions that can help foster a more constructive exchange:

**How do you balance competing perspectives and priorities regarding energy policy? What trade-offs do you consider when it comes to local, national, and global effects?**

Acknowledging that policymakers face difficult trade-offs and must weigh multiple competing interests is critical to building trust and rapport. This question invites a candid discussion about the various factors influencing a policymaker's decisions, from economic impacts to environmental concerns to social equity considerations.

**What data do you trust and incorporate into your decision making? Where do you source this data from?**

This question assumes policymakers use data, which paints you as a generous audience. It allows policymakers the chance to share and show off their expertise and insider knowledge. Also, the question(s) could open a discussion on the credibility and reliability of different data sources and how we can all identify areas where better information or more rigorous analysis is needed.

**How do you think about the impacts of energy regulations on the energy companies producing the power and the ratepayers, both businesses and ordinary citizens?**

This question drills down further into local effects, particularly local energy grid demands and the knock-on effects for ratepayers of all kinds. This knowledge can enlighten you regarding which groups the policymaker is most concerned about when weighing policy options.

The "how" is an important part of this question. Asking "how" requests the policymaker explain their thinking process and opens the door for you to have a discussion on other possible ways of thinking about the impacts. (This is compared to asking "what" which will result in an opinion with no context.)

**What do you see as the role of the government in driving**

**innovation and deployment of energy efficiency technologies? What former policies or investments did you think would be most effective, and which actually created change?**

Knowing where they lie on government involvement or guidance in economic sectors is essential to collaborate with a policymaker. Asking about a policymaker's vision for the government's role in the energy transition can reveal important insights into their underlying philosophy and approach to regulation—which will range from removing all regulation to setting strict guardrails for the energy sector. The second question is a necessary follow-up to help you clarify a dialogue about what is truly effective and what sounds good on paper. (If this question is part of a more extensive discussion, it may be relevant to bring answers and conclusions drawn from Question 2, above.)

**How do you think about the global impacts of energy policy, particularly regarding issues like international trade, diplomacy, and national security?**

Energy policy doesn't exist in a vacuum, and this chunky question acknowledges it. Municipal policymakers might be oblivious to how their local economy affects national and global economies. This question can start a conversation with them about the greater knock-on effects of stringent regulation. For federal and international policymakers, this question gives them a chance to expound upon the many factors impacting their decisions. As the listener, you better understand how much the policymaker weighs global effects into their decisions and which aspects (economics, diplomacy, or national security) they consider most.

**How do you engage with constituents and stakeholders who disagree with your positions on energy issues? What strategies have you found effective for finding common ground and building consensus?**

Asking about a policymaker's approach to engaging with critics and opponents can provide valuable insights into their leadership style and conflict resolution skills. The second question encourages the policymaker to focus on successful collaboration rather than dwelling on instances of disagreements (which would only stagnate a pragmatic discussion about solutions).

**What opportunities do you see for collaboration between policymakers at different levels of government?**

Policymakers don't act in a bubble—many are affiliated with a political party and other policymakers. This question gauges their involvement with the policymakers above, below, and lateral to themselves. It allows you to understand how connected they are and how much power they may wield when pushing regulation forward.

**What do you see as the most significant barriers to progress on clean energy and climate policy, and what strategies do you think are most promising for overcoming those barriers?**

Asking policymakers to identify obstacles and opportunities is a great way to understand their perspectives on challenges and stimulate conversation about solutions. Policymakers could easily avoid action because the barriers are too high or comprehensive, but the second part of this question encourages them to push forward over and around barriers.

**How do you balance the need for long-term planning and investment in the energy system with the short-term pressures of election cycles and political expediency?**

This question gets at the heart of one of the biggest challenges facing energy policymakers: the mismatch between the timescales of energy infrastructure and the political process. It invites a candid discussion about how policymakers think about and navigate these competing pressures. Do they truly work to build durable solutions that can withstand administration changes and shifting political winds? Or are they more concerned about short-term political gains? Answers to this question often reveal how dedicated a policymaker is to their community.

. . .

By asking questions like these, we can gain a deeper understanding of the motivations, constraints, and priorities shaping policymakers' decisions on energy. However, even armed with good questions, policymakers can be tricky to communicate with. They differ from the energy companies and environmentalists in that they are usually expert public speakers. Some are careful with their words to the point of avoiding commitment to any one viewpoint. Many will give non-answers that don't please anyone, but also don't displease anyone. Because they are expert evaders, it might feel like policymakers ignore questions. Usually, it is not an intentional dig; instead, the policymaker is often uncomfortable and so avoids answering a question. To this end, calling out policymakers when they avoid answering inquiries will be essential.

But, like asking good questions, calling out someone on their evasiveness should put the answerer at ease rather than on defense. When a policymaker avoids or ignores a question, it's important to approach the situation with empathy while firmly pursuing an answer. Most of them are concerned about public perception and will be afraid to veer from repeated stances on issues. You can try a few of the following responses to draw an answer out of them while remaining empathetic to their position:

- "I understand you might not want to engage in this dialogue—it's a complex issue needing more attention than the confines of our timed discussion today. We'd really love to hear your high-level answer to the question..."
- "I didn't ask this question because I was seeking a commitment to action from you or your affiliates. I just want to hear your perspective on this issue.

Given your position and expertise, what is your opinion on…"
- "Party lines aside, I'd love your individual opinion. I realize it might not be something you talk about often. Would you like to ask me any clarifying questions before you answer?"

Remember, the goal is to create an environment where the policymaker feels comfortable sharing their honest thoughts while holding them accountable to address your questions. Acknowledging another person's discomfort can encourage them to speak. A policymaker may feel more willing to dialogue once the uncomfortable nature is addressed. (There are a select few policymakers who will resent you for pointing out that they seem uncomfortable with specific lines of questioning.)

With good questions and some language to further prod policymakers, we'll explore strategies and success stories for collaborating with policymakers to drive change in the energy sector.

## COLLABORATING WITH POLICYMAKERS

When working with policymakers, it's crucial to temper and manage expectations. You must understand at the outset that they are trying to balance the needs and interests of a wide range of constituents, from individual voters to campaign donors to their party. Plus, they need to factor in the effects of policy and regulation at the local, national, and global levels, both in the short-term and long-term. Energy is fundamental to everyday life, so any regulation policymakers enact will have ripple effects across the economy and society. Policymakers must weigh the impacts on jobs, energy prices, reliability, the environment, and more.

At the same time, policymakers operate within the constraints of election cycles. This can create pressure to prioritize short-term wins over long-term planning. They must constantly be attuned to how the public and the media will perceive their actions. And they must navigate a complex landscape of existing laws, regulations, and jurisdictional boundaries. Given all these realities, policymakers are perhaps the most difficult of the three groups to collaborate with.

Policymakers can sometimes seem risk-averse or slow to embrace change, even when they recognize the need for action. They are understandably wary of balancing actions leading to positive progress and actions seen as overstepping mandates. The most effective collaborations with policymakers consider these challenges and look for ways to create wins for policymakers in the eyes of all their constituents.

One promising approach is to focus on energy efficiency as a unifying goal. Improving efficiency—getting more economic output and social benefit from each unit of energy consumed—has the potential to address multiple policy objectives simultaneously. It can lower costs for consumers and businesses, improve reliability by reducing strain on the system, enhance competitiveness by driving innovation and productivity, and deliver environmental benefits by cutting waste and emissions.

Importantly, energy efficiency is not an ideological or partisan issue. It appeals to conservatives as a way to promote free market solutions and reduce the need for heavy-handed regulations. It appeals to liberals as a tool for environmental protection and consumer empowerment. And it appeals to ratepayers because energy efficiency means a lower monthly bill. By framing efficiency as a shared priority, it's possible to bring together unusual bedfellows and build broad-based political support.

Before anyone says it's impossible to meet with policymakers in the Radical Middle, let's consider a few examples of an energy-efficiency-first (aka Radical Middle) approach. One such example is the EmPOWER Maryland program, which was first established in 2008 and has been extended and expanded several times with strong bipartisan support. EmPOWER set ambitious targets for the state's utilities to achieve annual energy savings through customer efficiency programs, with incentives for exceeding the goals and penalties for falling short.

The key to EmPOWER's success was its collaborative and iterative design process. Policymakers worked closely with utilities, environmental groups, consumer advocates, and other stakeholders to craft a framework that balanced interests. Utilities received flexibility and cost recovery for their investments. Consumers received access to valuable rebates and services. Environmentalists received measurable emissions reductions. The state economy received a boost from the jobs and economic activity generated by the efficiency industry.

Another example comes from the US Southeast region, where several utilities have partnered with state and local governments to launch energy efficiency financing programs for low and moderate-income households. These programs provide long-term, low-cost loans that are paid back through the energy savings on customers' bills, making efficiency upgrades accessible to families who might not otherwise be able to afford them. Recognizable names from the program include Energy Star and the "Pay As You Save®" (PAYS®) program.[4]

Again, the key has been finding a goal that works for all constituents. There are no tax penalties on mid-sized energy producers for being unable to decrease emissions—instead, there's support to help them better their operations. Energy costs decrease as power plants and ratepayers try to reduce their usage in tandem. (This has the added benefit of opening up the power supply for businesses' increasing power needs.) By focusing on co-benefits and shared interests, these programs have been able to overcome traditional barriers and drive significant investment into underserved communities.

As these examples show, collaborating effectively with policymakers requires a willingness to put yourself in their shoes and look for pragmatic, politically viable solutions. It means being clear-eyed about their constraints and recognizing the unique levers they possess to drive systemic change. Most of all, it means leading with shared goals and values.

Ultimately, the measure of success for energy policy is real-world impact. A Radical Middle approach brings policymakers to the table, drawing them toward real action over mere talk of making a difference. Policymakers are vital in setting the rules and creating incentives of our energy future. But they can't (and, let's be honest, shouldn't) do it alone, and they definitely can't do it from the extreme ends of the ideological spectrum. Lasting progress will require sustained collaboration and compromise from all of us. Leading with quantitative facts, incorporating curiosity and empathy, and focusing on shared interests is an excellent place to start.

But even when we've stayed curious, asked the right questions, and appear to be on the path to pragmatic action, we might stumble into a few more barriers, not of our own making. We'll dive into those (and how to avoid them) next.

# CHAPTER 9
# STAYING THE COURSE WHEN OPPOSITION MOUNTS

Finding your way to the Radical Middle is not easy. Curiosity aside, our society tends to penalize people who do not remain anchored to a side. There will always be detractors and people who are so afraid to be wrong they will never compromise. The incentive structure dictating our political system does not encourage curiosity—it endorses polarization. It's filtered down to interactions in businesses and schools. Even our media has ventured off the path of conventional news and is now seemingly obsessed with political rhetoric. This has resulted in difficulties with regular polite conversation. I know this firsthand from countless frustrating interactions I've had over the years.

I remember one particularly exasperating exchange at an energy conference a few years back. I had just given a presentation on the importance of collaboration and working toward energy efficiency goals. Afterward, an environmental advocate approached me with a frown on her face. I vaguely recognized her as a career woman in her field.

"I appreciate what you're trying to do," she said, "but I just don't see how we can play nice with the fossil fuel industry. Getting along with each other is simply never going to happen.

They've been pulling the wool over our eyes and throwing up roadblocks for ages. Why should we start trusting them now?"

I could've just brushed her off, but something about the fire in her eyes made me pause. Her question was laced with passive-aggressiveness—it was not a good question. That said, I had a hunch that under the confrontational tone, there was a real passion for making the world better. If I could just hash out where she was coming from, maybe we could start chipping away at the barrier between us.

"I hear your frustration," I said. "And you're right, the fossil fuel industry has a lot to answer for. But the reality is, like it or not, they're still a powerful force in energy, in our economy—and our country's overall well-being will remain reliant on fossil fuels over the foreseeable future. We can't just wish them away—and I am not so sure detractors, let alone the general public, fully understands just how extensive our reliance on fossil fuels actually is. If we want to make real progress on the issues you're passionate about, we need to find ways to get them on board and work with them."

"I don't see how that's possible," she said. "Their whole business model is based on extracting and burning dirty fuels. They're not just going to change overnight."

"It's true any big company will be slower to change," I said. "But what if we could find ways to align all our interests? What if we could create incentives for them to invest in clean energy and carbon reduction technologies instead of disproportionately penalizing them for emissions?"

"I don't buy it," she said without a second's pause. "They'll just use those incentives as a fig leaf to keep polluting. We need to completely phase out fossil fuels, not help them greenwash their image."

I tried one last time to find a point of connection.

"I understand your skepticism," I said. "But, even if we could snap our fingers and make fossil fuels disappear tomorrow, we'd still need to figure out how to power our economy and keep everyone's lights on. We've been ramping up renewables for years and they're just not built for scale yet—and they're not efficient enough for the US's power needs, let alone the world—not to mention they remain expensive and unreliable. We need an all-of-the-above approach that includes both renewable energy and energy-efficient fossil fuels. Isn't it better to have a seat at the table and try to steer things in the right direction?"

"Thanks for hearing me out," she said, "but I think we're just going to have to agree to disagree on this one."

As she walked away, I felt a mix of disappointment and confusion. I typically feel a bit crazy after having these conversations. I am not sure why, but I have a tendency to second guess my stance on these topics, even though I've done the analysis probably one hundred times. I had hoped that by engaging with her directly and trying to understand her perspective, I could start to build a bridge between our two worlds—or at the very least, get some degree of buy-in on what I think are a practical set of solutions. But in the end, like many others, she wasn't ready to engage. And she departed before I even had a chance to bring up nuclear power as an option.

As frustrating as these moments are, they're par for the course on the road to the Radical Middle. Some people will refuse to engage no matter how hard you try. The thought of even entertaining constructive dialogue regarding fossil fuels is an impossibility. I've had many more people—from lawyers, to academics, to fossil fuel advocates—give me the brush-off. In these scenarios, it is often better to let the person walk away. Suffice it to say they aren't ready for the Radical Middle yet, and you'll have to try again next time.

To make progress toward meeting in the Radical Middle, though, it's not just about chipping away slowly at those who are most entrenched in their ideologies. There are several other factors that can mitigate a person's unwillingness to meet in the Radical Middle. One of the biggest factors I see today is the widening generational divide between incumbent leaders and younger stakeholders clamoring for faster, smarter change.

## THE GENERATION GAP DIVIDING US

Generational divides don't usually get much media attention, but they definitely impact the energy space—to the point where it sometimes seems the debate is not pro- versus anti-fossil fuel, but Boomer versus Millennial. Bridging the gap between incumbent industry players and younger stakeholders is critical for unlocking new forms of collaboration and accelerating change in the energy sector.

Generational divides are difficult to navigate, as they are rooted in deep-seated differences in values, life experiences, and communication styles. The woman who approached me perhaps had more riding on her hardline beliefs because she has held them for thirty years and clearly made a career out of it. In the industry, we tend to refer to this dynamic as "Stubborn Dad Syndrome." The generational dynamic is akin to me trying to get my dad to change how he goes about his day—which is to say, I try to persuade him, and he resists. He won't change—it's just not going to happen.

In the energy space, these divides play out in multiple ways. Within companies, older leaders who have spent their careers in the fossil fuel industry may be resistant to change, seeing calls for a rapid transition as naïve or unrealistic. They may dismiss younger employees pushing for more aggressive climate action as idealistic, unemployable troublemakers who don't understand how business works. This dynamic can stifle innovation and drive away top young talent who could help a corporation diversify risk and be better prepared for the future.

For instance, the nuclear power debate is a prime example of how generational divides can hinder progress. Many older leaders in energy and energy-adjacent industries vividly remember the nuclear accidents at Three Mile Island and Chernobyl. These events were each terrifying at the time, and media coverage stoked widespread public fear. Understandably, a general opposition to nuclear energy arose over the intervening years.

The generations who have Chernobyl in living memory are much more likely to oppose nuclear energy, despite how safe it has become in the thirty-plus years since Chernobyl. As mentioned in an earlier Chapter, the state of Illinois is majority (67 percent) powered by nuclear energy and produces so much energy that it sells excess power to surrounding states. The US Navy, too, largely runs off nuclear power—it is, in fact, a military technology and has proven itself to be an incredibly safe source of power in 2024. Yet most people over the age of 60—many of whom dictate policy—are opposed to incorporating more nuclear power into the energy mix. And even the ones who do endorse nuclear tend to argue against it when it is presented in the context of a political debate.

Meanwhile, younger generations seem less fearful of nuclear power—many younger people I have spoken to think nuclear energy is promising. To them, Chernobyl is a freak accident discussed in history textbooks and cable miniseries. Three Mile Island was a near miss. Most people under the age of 60 embrace the benefits of nuclear energy as a solution to their climate concerns, but also as a way to have cheaper energy as demand increases. The nuclear energy debate is just one area where disagreements rooted in generational differences lead to missed opportunities.

The "old guard" currently holds almost all of the sway in the energy sector. CEOs, board members, and policymakers, even environmentalists, who built their careers in a different era can be set in their ways and resistant to new ideas. This is not entirely negative. The old guard has decades of experience and hard-earned knowhow, and they boast helpful battle scars. But the generational gap harms progress when established leaders overestimate their own importance and knowledge, and become out of touch with the times. This overconfidence blinds them to coming issues (like the skyrocketing demand for energy) and innovation that could improve outcomes. They stick with what they know and tend to ignore other perspectives.

At the core of generation issues, there is often a failure of curiosity. Many veteran industry players become "know-it-alls" and are more inclined toward making snap judgments than to exploring novel possibilities. They think the way things have always been done is the way they must continue. They see deviations from the well-trodden path as too risky or not worth their time. And sure, some of the proposals presented by younger generations are asinine—yet, many ideas from the younger generations have merit. Younger generations have come of age in a time of technology where disruption is the norm and have only survived by embracing change and adapting. The energy sector needs more of this attitude if it intends to keep up with the coming spike in demand.

The old guard must recognize the industry has shifted—the pressing realities of energy security, technological advancement, and market demands have rewritten the rules of the road. Simply tweaking the edges of business-as-usual is no longer enough. That does not mean the incumbent generation cannot remain as mentors, only that their time in office, so to speak, is at an end.

Seizing the full potential of energy efficiency will require a generational shift in leadership. The old guard must transition into advisory roles for the incoming leaders and relinquish their seats at the table. This is not just about training their replacements but about knowing when to cede the stage and create openings for new leaders to emerge. For all the controversy and headaches associated with the younger generation, the truth is that the Millennials are no longer kids—they are now over forty years old. Gen Xers are in their fifties, and Gen Z is quickly approaching 30.

However, even when we solve for generational divides, there is still another problem I often encounter when trying to approach the Radical Middle with others: the acronym ESG—the factors we've discussed throughout this book affecting energy—is a dirty word.

## ESG IS A DIRTY WORD

The polarized debate around Environmental, Social, and Governance (ESG) investing highlights the challenges of advancing goals in a politically charged environment. ESG is meant to be a better way to understand economic impacts, yet so many people judge me based on the "ESG-expert" label. To be honest, I would rather be called an asshole. The "ESG" label drives me nuts. Both ends of the ideological spectrum have disparaged the work I do on issues considered ESG-related.

The practical application of ESG implies the evaluation of non-financial factors influencing margin, cash flow, and the long-term cost of capital. Advocates of extreme ends of the ideological spectrum see the data from ESG as a threat to the status quo. And it is.

In recent years, ESG has become a lightning rod, with critics on both the right and the left assailing it for very different reasons. On the right, ESG is often portrayed as a form of "woke capitalism" undermining shareholder value and free market principles in pursuit of an environmental agenda. Conservative politicians and think tanks have launched concerted attacks on ESG, pressuring companies and investors to abandon sustainability metrics and focus solely on financial returns. Texas has gone so far as to ban its state pension funds from doing business with financial institutions that "boycott" fossil fuels based on ESG criteria.

Meanwhile, on the left, some activists dismiss ESG as corporate greenwashing. These activities suggest ESG fails to drive meaningful change. They argue that voluntary ESG reporting and investing allows companies to burnish their reputations without fundamentally altering destructive business practices. In their view, only government regulation, legal action, and public pressure can force the drastic emissions reductions needed to avert climate catastrophe.

Caught in the crossfire are the investors, companies, and advocates who see ESG as a vital data management tool for managing risks, seizing opportunities, and aligning the capital markets with the needs of a sustainable future. For them, considering environmental and social factors is not about sacrificing returns, but about generating long-term value and resilience. Ironically, the energy space always evaluated environmental, social, and governance characteristics; they just never formally codified it. They point to the growing body of evidence linking strong ESG performance with better financial outcomes, from lower costs of capital to higher profitability and stock returns.

Critics conveniently ignore the fact that ESG grew out of a recognition by mainstream investors that long-term financial success is inextricably linked to maintainable environmental and social outcomes. For well over a decade, large asset managers, banks, and financial institutions have used ESG data and analytics to gain insights into corporate performance and make more informed investment decisions. This isn't about politics or ideology, and frankly, it should have never gotten to this point. ESG is about objective data and *thorough* empirical analysis.

Companies managing environmental risks, treat communities (including employees) well, and have robust governance tend to outperform over the long run. The math is simple—consistently doing the right thing generally results in positive results. Conversely, those that pollute recklessly, exploit stakeholders, or engage in unethical practices often see their value erode due to regulation, litigation, reputational damage, or operational disruptions.

In today's hyper-partisan climate, even discussing ESG in a factual, practical, data-driven way has become fraught. When I do presentations and talks, I get hit from both sides of the aisle. The folks hurling accusations (which range from "anti-climate" to "woke") seem incapable of understanding that the data I present is not "Dan's Proprietary Data" set. In fact, most of the data I use can be found in public reports from government agencies—anyone can access it.

The fraught discussions make it incredibly difficult to have productive conversations about how to actually use ESG-related tools and strategies to drive positive impact. At my company, we've refrained from using ESG as a term to describe what we do. Instead, we're calling our line of work "Consultancy & Advocacy" because that's the essence of what we do. We use data to forecast trends, stay atop of policy, and help energy companies, policymakers, and environmentalists understand what is coming down the pipeline, and how to prepare for it.

I hope readers of this book better understand ESG, yet I'm inclined to abandon the term altogether. It has served its purpose, and now we need a new term for what ESG folks actually do because it is important. Sometimes, it feels like I'm the only one looking at the data and realizing the coming trends.

When you engage with a person on approaching the Radical Middle, throwing "ESG" into your word soup often backfires. Instead, advocate for looking more closely at all the data (from reputable, unbiased sources) and for integrating non-obvious non-financial factors into economic analysis to enable smarter, more holistic decision making. By grounding the conversation in facts and figures, ESG shifts power away from the loudest voices and toward those with the most credible solutions.

## STAYING FOCUSED ON THE RADICAL MIDDLE

With global energy demand set to go through the roof in the coming years, we need all hands on deck to make sure the transition is both fast and fair. To that end, the emphasis of energy talks should center on implementing thoughtful changes quickly.

It's incredibly easy to get pulled into the drama when confronted with political polarization and ideological extremism. But if we want to move the needle toward pragmatic action, we need to stay focused on maintaining a Radical Middle Mindset. Maintain curiosity in the face of big claims: What data is the other person using? What factors are they weighing the most heavily? How might your solution benefit the other person?

The energy efficiency focus of the Radical Middle approach is often the secret sauce for getting everyone on the same track. By homing in on getting more bang for our buck with every unit of energy, we can tackle three challenges at once: expanding affordable energy access, reducing the impact on the environment, and maintaining revenue for energy producers and suppliers. Energy efficiency is a win-win-win. For ratepayers, it means lower energy bills and more money in their pockets. For policymakers, it means a clear goal for creating policies. For utilities and grid operators, it means less stress on the system and more wiggle room to bring in innovation. And for all of us, it means cleaner air, water, and land.

But if we want to see these benefits on a larger scale, we need everyone in the energy space to collaborate. Divestment agendas get us nowhere. Skewing the fossil fuel narrative may provide some short-term perks, but it adversely impacts long-term prosperity. Ignoring non-financial performance negatively impacts long-term performance and cost of capital. We're talking about policymakers teaming up with fossil fuel companies to better understand their needs. Environmentalists and utilities joining forces to create efficiency programs that don't leave anyone behind. And all three groups—energy companies, environmentalists, and policymakers—respond not just to local energy needs, but national and global ripple effects. This is not a pipedream: detractor voices are not large in numbers—they are just annoyingly louder than others.

When people get high and mighty with moral arguments, Radical Middle advocates need to be ready to bring it back to the numbers. How much cash are people saving thanks to efficiency investments? How many jobs are being created in the clean energy sector? What are the health perks of cutting down on local air pollution? By always steering the debate back to objective data and realistic results, it gets harder for the extremists on either side to derail the conversation.

The Radical Middle is a marathon, not a sprint. It takes patience and persistence (and, for me, sometimes bourbon). Not everyone will be ready to compromise immediately. Most people will need several exposures to the Radical Middle Approach before they buy into it. Progress might be slow at times, but if we keep chipping away one conversation at a time, we can all meet in the Radical Middle.

We need to solve for energy locally, nationally and globally. And ideally, we do so before another global power does.

# CHAPTER 10
# AN "ALL OF THE ABOVE" SET OF ENERGY SOLUTIONS WILL MAKE THE WORLD A BETTER PLACE

Energy affordability and standard of living are inextricably linked—this is not a matter of opinion; it is mathematically correlated. For far too many Americans (and citizens globally), high energy costs are a significant barrier to financial stability and upward mobility. Nearly four of five U.S. households live paycheck to paycheck. Energy affordability and upward mobility has essentially remained static for the last quarter century. Families struggling to make ends meet face agonizing trade-offs between keeping the lights on and putting food on the table.

Unfortunately, this is not an exaggeration. Energy insecurity not only causes immense personal hardship, but also stifles economic growth and community development. By making energy efficiency and innovation a top priority, we can break this negative cycle and raise the standard of living for millions of households currently underserved by our energy systems. Fortunately, the Radical Middle approach provides a pragmatic blueprint to overcome these burdens.

Consider the hypothetical situation of Hank, a single dad living in an energy insecure county in rural America. He works two jobs to support his young daughter. Every month is a struggle to keep up with the bills, especially in the cold winter months when the aging furnace in his modest rental home must work overtime.

Some months, Hank must make the excruciating choice between keeping the heat on and putting enough food on the table. He's fallen behind on the utility bills more than once, forcing him to turn to assistance programs and take on debt just to keep the lights on and his little girl warm. It's a stressful, precarious existence—one missed shift or unexpected expense could mean not making rent.

Hank's story is all too common in his community, where energy costs consume a disproportionate share of household incomes. In his county, the energy infrastructure is outdated and inefficient, with power outages during peak usage (think the very coldest winter days, and hottest summer months) and price spikes. The local utilities have been slow to invest in upgrades for a variety of reasons, though a major one is spending capital on tax penalties for emissions.

But then, local policymakers work to better understand the needs of the local power plant. Through their discussions, they collaborate on solutions to help the utility improve efficiency. An environmentalist group is brought in to help with the effort and reduce emissions, so the energy company doesn't have to devote its capital to the tax penalty.

As these efforts scale up, Hank starts to see his utility bills stabilize and even decrease, despite the harsh winters and scalding summers. With the modest energy savings, he's able to put a little aside for emergencies and enroll his daughter in an after-school soccer program. He breathes a little easier knowing he can keep the heat on without sacrificing essentials and provide opportunities for his daughter.

The benefits ripple out to the broader community too. The local public school has lower energy bill and can afford new books. As energy efficiency improvements free up capacity on the grid, the county becomes more attractive for new business investment. A data center company decides to build a new facility in the area, drawn by the reliable, affordable, and innovative power supply. The project creates hundreds of jobs and increases demand for more energy efficiency improvement efforts.

Hank's story demonstrates how collaboration on energy efficiency could translate into tangible improvements in people's lives. A 2022 report found more than 20 million U.S. homes are behind on their energy bills.[1] That is one of every six U.S. households, the highest number on record. For families living on the margins, even small reductions in energy burdens can be life-changing.

Moreover, when citizens' savings are invested back into local communities, they can spark improving cycles of economic development and opportunity. "Local" is the key here. Federal policy is increasingly ineffective. Policy can act as a referee, but it certainly should not be a player dictating local and regional energy actions. What works in Florida probably does not make sense in Illinois, and what works in Texas probably does not make much sense in Maine.

The Radical Middle strategy isn't just a way to improve lives domestically in the US though. The Radical Middle approach is vital to preserving the American way of life. How the U.S. navigates the changing energy market will have profound implications for our country's standing on the world stage and the strength of democracy globally. For every piece of U.S. energy, there are extensive global repercussions. Getting it right is about more than keeping lights on in homes—it's about harnessing the power of energy to preserve the American way of life for generations to come.

## THE FUTURE OF THE FREE WORLD IN QUESTION

The energy challenges facing the United States and the world are not just daunting in scale and complexity—they are also deeply interconnected with the fabric of our economy, our security, and our way of life. The coming surge in global energy demand is not a distant or abstract threat, but an urgent and all-encompassing reality that will test the limits of our ingenuity and resilience. How the US responds to the challenge will also largely determine our lasting position in the global geopolitical landscape over the next century.

Consider this: the United States alone will need to generate enough additional electricity to power the equivalent of five Tokyos just to keep pace with the growing demands of the digital economy. That's a staggering amount of energy, and it's just the tip of the iceberg. Globally, energy consumption is projected to increase by nearly 50 percent by 2050, driven largely by population growth and rising living standards in developing countries.

Meeting this surge in demand will require a massive mobilization of resources and talent on a scale not seen since the industrial revolution. It will mean fundamentally reimagining energy systems, from the way we generate and distribute power to the way we consume and conserve it. And to be successful, it will require cooperation and collaboration across sectors and ideologies—we need the Radical Middle.

Nowhere is this more evident than in the battle over critical energy technologies and the digital infrastructure they power. The AI revolution already transforming our economy will be largely shaped by the countries that can harness the immense computing power and data flows it requires. This digital transformation is also enormously energy intensive and is the cause for predictions of the 50 percent increase in energy consumption by 2050.

As Chief Brody says in the legendary 1975 horror film *Jaws*: "We're gonna need a bigger boat." The coming demand for more energy is the shark chasing down and threatening to cripple (if not kill) the US, if we don't get in a bigger boat. We need to replace Chief Brody's broken-down fishing boat from the 1970s with a brand-new Andros Tarpon 26—which is to say, a better, bigger boat.

Frankly, the United States is currently underprepared for the impending global economic landscape. Existing and proposed future policies threaten to undermine our country's ability to remain the dominant force within the global economy. It seems many policymakers specifically overlook the energy impacts from the growth of the digital economy and initiatives to reshore vital industries.

This is where the nexus of energy efficiency and technological dominance becomes so crucial. Countries mastering clean, efficient, and affordable energy solutions will not only reap economic rewards, but also gain a decisive advantage in the generative AI arms race. Conversely, countries with uncertain energy futures risk being left behind in the digital dust.

China, in particular, has made no secret of its ambitions to dominate both the energy and digital domains. Through its "Made in China 2025" initiative and other industrial policies, Beijing has poured billions into cornering the market on everything from solar panels and batteries to semiconductors and 5G networks.[2] It is now leveraging this position to gain a stranglehold on the digital supply chains underpinning the global tech economy.

Consider the example of rare earth elements, which are essential for everything from wind turbines and EVs to smartphones and defense systems. China now controls over 85 to 90 percent of the global supply of these critical minerals (either by mining, refining, or distribution), giving the country enormous leverage over digital transitions and the clean energy sector.[3,4] China has not hesitated to use this power to extract geopolitical concessions and punish countries challenging its interests.

But China's ambitions go beyond just cornering markets. It is also using its growing technological prowess to export its model of authoritarian surveillance and control. Through its "Digital Silk Road" initiative, China is building digital infrastructure in countries around the world, from 5G networks in Europe to smart cities in Africa. These projects not only give Beijing access to vast troves of data, but also allow it to shape the norms and standards of the digital age in its own image.

If the U.S. fails to counter China's techno-authoritarian ambitions, it risks ceding control of the industries of the future—from clean energy to artificial intelligence—to a regime that does not share our values or interests. The consequences for American competitiveness, security, and freedom could be devastating.

But the U.S. is not powerless in the face of this challenge. The U.S. is a leader in current economic markets including *energy*. While China's citizens choke on fumes from the constant churn of dirty coal power plants, the U.S. is pioneering energy efficiency with fossil fuels. The U.S. has become energy efficient enough to export fuel sources like natural gas, oil, and coal to other nations, building wealth and political clout.

The stakes could not be higher—for the competitiveness of our economy, the security of our democracy, and the future of the free world. The U.S. cannot afford to be complacent or divided. Above all, it cannot fall into the trap of allowing ideological rhetoric to dictate policy. We cannot allow pragmaticism, reason, and economics to take a back seat to narcissistic dogma. The United States needs a national strategy that puts energy innovation and digital leadership at the center of its domestic and foreign policy agenda.

The Radical Middle approach engages pragmatically with the world as it is, while shaping the world toward what it could be. It recognizes that the path to a freer, more prosperous digital future runs through the complex realities of energy—and progress requires forging new collaborations and championing shared values. The Radical Middle provides a blueprint for a vision for an American future that is not only richer and more efficient, but also freer and more secure.

## THE TIME FOR THE RADICAL MIDDLE IS NOW

We don't have the luxury of debating whether or not to adopt a Radical Middle approach. The energy transition is already underway, and the stakes could not be higher. The countries that master energy technology and increase capacity will not only reap the economic benefits but also shape the geopolitical balance of power for generations to come.

In this book I have argued there is no one energy source that can power the grid on its own. Just like a stock portfolio should be diversified, so too should our energy mix. If the price of oil skyrockets, it's good to have the option to ramp up natural gas power or harness hydroelectric power. It's safer and more profitable to have a mix of energy sources to power our grids. In the good times we can export excess, and in the bad times we can utilize more of our own resources.

The energy makeup of America is not a matter of ideology or preference—it is a matter of physics and economics. Even if we could make fossil fuels disappear tomorrow, we would still face enormous technical and logistical barriers to meeting our energy needs with renewables alone. The intermittency of wind and solar power, the limitations of current battery storage technologies, and the sheer amount of land and materials required to scale up these systems to the level needed is staggering.

Moreover, by abandoning oil and gas prematurely, we would be handing a major strategic advantage to other competitors (for example, the nations that make up OPEC) who are only too happy to fill the void. We would be sacrificing the energy security and economic leverage that comes with being a major producer and exporter of these fuels, while also ceding the innovation potential of the oil and gas industry to develop cleaner and more efficient technologies.

Instead, we need a balanced, pragmatic approach that recognizes the ongoing role of fossil fuels in the energy mix, while also accelerating innovation for cleaner and more sustainable practices. This means lifting tax penalties on fossil fuel companies (which largely hurt the Middle Guys—small and mid-sized producers—not the big fossil fuel corporations), and instead allowing those companies to invest in innovations to increase energy efficiency.

This will mean setting aside the ideological battles and coming together around a shared vision of a cleaner, more secure, and more prosperous energy future. It means recognizing the challenges we face are too big and too complex to be solved by any one faction or worldview, and progress will require compromise, collaboration, and a willingness to adapt to changing realities. It also means the world must face the fact that the reliance on fossil fuels will remain in place over the foreseeable future, so the focus must remain fixated on efficiencies and technology—not replacement and elimination.

The Radical Middle approach offers a path forward, one rooted in science, economics, and pragmatism rather than wishful thinking or political posturing. It recognizes that adapting to our energy needs will require investment, and leadership over the course of decades.

We have no time to waste. The clock is ticking, and the stakes are rising with every passing day. China and other competitors like Russia and Saudi Arabia are not waiting for us to get our act together—they are moving aggressively to shape the geopolitical landscape in their favor. To combat their actions, we need to mobilize every tool at our disposal—from cutting-edge research and development to smart regulation and diplomacy—to accelerate energy efficiency and secure our place in the industries of the future. And to do so, we'll need to collaborate with each other, using the full force of our collective knowledge and efforts. Like it or not, Americans will all need to meet in the Radical Middle.

As we come to a close, I'm reminded of where we started in Chapter 1: with the recognition that energy is the lifeblood of modern civilization. It powers our economy, our security, and our aspirations for a better world. The great task ahead is to evolve our energy systems to meet the needs of the 21st century and beyond. A Radical Middle Mindset and meeting others in the Radical Middle equips us to navigate this transition with ingenuity, pragmatism, and a commitment to collaborating on all sorts of solutions that work for all.

# NOTES

## 1. GREEN AIN'T CLEAN


1. Ivan Penn, "California's Electric Grid Has a Wild New Enemy: Heat," *The New York Times*, September 25, 2022, https://www.nytimes.com/2022/09/25/business/energy-environment/california-energy-grid-heat.html.
2. U.S. Energy Information Administration, "Washington State Profile and Energy Estimates," *U.S. Energy Information Administration*, accessed January 10, 2025, https://www.eia.gov/state/print.php?sid=WA.
3. U.S. EIA, U.S. Nuclear Generation and Generating Capacity, 2023P.
4. U.S. EIA, Electricity Data Browser, Net generation for all sectors (thousand megawatt hours), Illinois, Annual, 2001-22.
5. U.S. Energy Information Administration, "Electricity Data Browser," *U.S. Energy Information Administration*, accessed January 10, 2025, https://www.eia.gov/electricity/data/browser/#/topic/0.
6. U.S. Energy Information Administration, "Virginia State Profile and Energy Estimates," *U.S. Energy Information Administration*, accessed January 10, 2025, https://www.eia.gov/state/?sid=VA.
7. U.S. Energy Information Administration, "Virginia State Profile and Energy Estimates," *U.S. Energy Information Administration*, accessed January 10, 2025, https://www.eia.gov/state/print.php?sid=VA.
8. Ember (2024); Energy Institute - Statistical Review of World Energy (2024) – with major processing by Our World in Data. "Electricity generation from coal – Ember and Energy Institute" [dataset]. Ember, "Yearly Electricity Data"; Energy Institute, "Statistical Review of World Energy" [original data]. https://ourworldindata.org/grapher/electricity-prod-source-stacked?time=1996..latest
9. Evan Halper, "Amid Explosive Demand, America Is Running Out of Power," *The Washington Post*, March 7, 2024, https://www.washingtonpost.com/business/2024/03/07/ai-data-centers-power/.
10. Our World in Data, "Primary Energy Consumption," *Our World in Data*, accessed January 10, 2025, https://ourworldindata.org/grapher/primary-energy-cons.
11. International Energy Agency, *Electricity Market Report 2023*, accessed January 10, 2025, https://iea.blob.core.windows.net/assets/255e9cba-da84-4681-8c1f-458ca1a3d9ca/ElectricityMarketReport2023.pdf.
12. U.S. Energy Information Administration, "Today in Energy: United States Remains Largest Producer of Petroleum and Natural Gas Hydrocarbons," *U.S. Energy Information Administration*, accessed January 10, 2025, https://www.eia.gov/todayinenergy/detail.php?id=3130.

# 2. THE VILIFICATION OF ENERGY COMPANIES

1. Associated Press, "California's Electric Grid Has a Wild New Enemy: Heat," *Los Angeles Times*, July 3, 1996, https://www.latimes.com/archives/la-xpm-1996-07-03-mn-20787-story.html.
2. David Heath, "How Industry Scientists Stalled Action on a Carcinogen," *Center for Public Integrity*, accessed January 10, 2025, https://publicintegrity.org/environment/how-industry-scientists-stalled-action-on-carcinogen/.
3. Simons, Marlise. "Will We Emerge in the 80's from the Dark Ages of Energy Policy?" *New York Times*, February 22, 1981. https://www.nytimes.com/1981/02/22/nyregion/will-we-emerge-in-the-80s-from-the-dark-ages-of-energy-policy.html.
4. "The Clean Air Act and the Economy." EPA, Environmental Protection Agency, 6 Jan. 2017, www.epa.gov/clean-air-act-overview/clean-air-act-and-economy
5. Hoffman, Andrew J. How Culture Shapes the Climate Change Debate. Stanford University Press, 2015.
6. Epstein, Alex. *The Moral Case for Fossil Fuels*. Portfolio/Penguin, 2014.
7. David Pilling, "How China Is Winning the Race for Africa's Lithium," *Financial Times*, January 10, 2023, https://www.ft.com/content/02d6f35d-e646-40f7-894c-ffcc6acd9b25.
8. Business & Human Rights Resource Centre, *China's Transnational Mining Operations: Human Rights Risks and Corporate Accountability*, accessed January 10, 2025, https://media.business-humanrights.org/media/documents/2023_China_TM_briefing.pdf.
9. Ashmoore, Olivia, et al. "Regional Carbon Dioxide Implications of the Retirement of the Navajo Generating Station." Energy Policy, vol. 134, 2019, p. 110981., doi:10.1016/j.enpol.2019.110981.
10. "Oil Sands Boreal Forest Conservation Initiative." Pembina Institute, www.pembina.org/project/oil-sands-boreal-forest-conservation-initiative
11. "REBA's 2020 Deal Tracker." REBA, rebuyers.org/deal-tracker/
12. Office of Governor Gavin Newsom, "Governor Newsom Signs Sweeping Climate Measures, Ushering in New Era of World-Leading Climate Action," *Office of Governor Gavin Newsom*, September 16, 2022, https://www.gov.ca.gov/2022/09/16/governor-newsom-signs-sweeping-climate-measures-ushering-in-new-era-of-world-leading-climate-action/.
13. New York State Senate, "Senate Bill S6599," *New York State Senate*, accessed January 10, 2025, https://www.nysenate.gov/legislation/bills/2019/S6599.
14. Carlos Nogueras Ramos, "Donald Trump's Promise to 'Drill, Baby, Drill' Probably Won't Change Much — Least of All in Texas," *The Texas Tribune*, August 15, 2024, https://www.texastribune.org/2024/08/15/donald-trump-energy-policy-fact-check-election-2024/.

## 3. US VS THEM DOESN'T WORK


1. International Energy Agency, "World Energy Outlook 2022 Shows the Global Energy Crisis Can Be a Historic Turning Point Towards a Cleaner and More Secure Future," *International Energy Agency*, accessed January 10, 2025, https://www.iea.org/news/world-energy-outlook-2022-shows-the-global-energy-crisis-can-be-a-historic-turning-point-towards-a-cleaner-and-more-secure-future.
2. U.S. Energy Information Administration, "Renewable Energy Provided 21% of U.S. Electricity in 2020," *U.S. Energy Information Administration*, accessed January 10, 2025, https://www.eia.gov/todayinenergy/detail.php?id=48296.
3. Rosie Frost, "Oil and Gas Projects Need Insurance – but Policies Are Being Pulled as Climate Change Bites," *Euronews Green*, October 19, 2022, https://www.euronews.com/green/2022/10/19/oil-and-gas-projects-need-insurance-but-policies-are-being-pulled-as-climate-change-bites.
4. Mia Wallace, "Revealed – Top 10 Fossil Fuel Insurers," *Insurance Business UK*, accessed January 10, 2025, https://www.insurancebusinessmag.-com/uk/news/environmental/revealed--top-10-fossil-fuel-insurers-465649.aspx.
5. Zachary Barlow, "Did BlackRock Abandon ESG?" *PracticalESG*, August 28, 2024, https://practicalesg.com/2024/08/did-blackrock-abandon-esg/.
6. Attracta Mooney, "Vanguard's Backing for Green and Social Proposals Falls to 2%," *Financial Times*, August 28, 2023, https://www.ft.com/content/4313afe4-1fee-447d-b05b-0c8c38cfbff1.
7. Chelsea Harvey, "China's Greenhouse Gas Emissions Exceed Those of All Other Developed Countries Combined," *Scientific American*, May 7, 2021, https://www.scientificamerican.com/article/chinas-greenhouse-gas-emissions-exceed-those-of-all-other-developed-countries-combined/.
8. Mengpin Ge, "Interactive Chart Shows Changes in the World's Top 10 Emitters," *World Resources Institute*, accessed January 10, 2025, https://www.wri.org/insights/interactive-chart-shows-changes-worlds-top-10-emitters.
9. European Commission, Joint Research Centre, *Emissions Database for Global Atmospheric Research (EDGAR) Report 2023*, accessed January 10, 2025, https://edgar.jrc.ec.europa.eu/report_2023#emissions_table.
10. CNN Staff, "See How the World's Biggest Polluters Are Changing – or Not – at COP28," *CNN*, December 2023, https://www.cnn.com/interactive/2023/12/us/countries-climate-change-emissions-cop28/.
11. United Nations Framework Convention on Climate Change (UNFCCC), "Kyoto Protocol," *United Nations Framework Convention on Climate Change*, accessed January 10, 2025, https://unfccc.int/kyoto_protocol.
12. "Canada Pulls Out of Kyoto Protocol," *CBC News*, December 12, 2011, https://www.cbc.ca/news/politics/canada-pulls-out-of-kyoto-protocol-1.999072.
13. Yunnan Chen, "Why Cobalt Mining in the DRC Needs Urgent Attention," *Council on Foreign Relations*, accessed January 10, 2025,

https://www.cfr.org/blog/why-cobalt-mining-drc-needs-urgent-attention.

14. CATAPA, "Lithium Exploitation Is Drying Out the World's Driest Desert," *CATAPA*, accessed January 10, 2025, https://catapa.be/en/lithium-exploitation-is-drying-out-the-worlds-driest-desert/.
15. Victoria Milko, Ed Davey, and Camille Fassett, "Indonesia's Massive Metals Build-out Is Felling the Forest for Batteries," *The Diplomat*, July 17, 2024, https://thediplomat.com/2024/07/indonesias-massive-metals-build-out-is-felling-the-forest-for-batteries/.
16. Charlie Hoffs, "Just and Sustainable Solutions for the Mining and Recycling of Rare Earth Elements in Wind Turbines," *Union of Concerned Scientists Blog*, accessed January 10, 2025, https://blog.ucsusa.org/charlie-hoffs/just-and-sustainable-solutions-for-the-mining-and-recycling-of-rare-earth-elements-in-wind-turbines/.
17. Bob Yirka, "Hydropower Dams Can Have a Big Impact on Local Climate Depending on Where They're Built," *Phys.org*, March 9, 2022, https://phys.org/news/2022-03-hydropower-impact-theyre-built.html.
18. Quirin Schiermeier, "Why Dams Are Razing River Ecosystems," *Nature*, January 3, 2020, https://www.nature.com/articles/d41586-020-00005-8.

## 4. THE RADICAL MIDDLE MINDSET

1. Bowen, Leann, writer. *Ted Lasso*. Season 1, Episode 8, "The Diamond Dogs." Directed by Declan Lowney. Released September 18, 2020.
2. Vermont Energy Investment Company https://www.veic.org/clients-results/case-studies/planning-the-future-of-energy-efficiency-programs-in-wisconsin
3. Jacob Bogage, "Data Centers Draw Scrutiny as They Multiply in Northern Virginia and Beyond," *The Washington Post*, April 30, 2024, https://www.washingtonpost.com/dc-md-va/2024/04/30/data-centers-regulations-northern-virginia-georgia-arizona/.
4. Virginia General Assembly, *Energy Plan 2021: Senate Document No. 17*, accessed January 10, 2025, https://rga.lis.virginia.gov/Published/2021/SD17/PDF.
5. American Exploration & Production Council, "Understanding the 2024 Methane Regulation Landscape," *AXPC*, January 4, 2024, https://axpc.org/2024/01/04/understanding-the-2024-methane-regulation-landscape/.
6. U.S. Energy Information Administration, "West Virginia State Profile and Energy Estimates," *U.S. Energy Information Administration*, accessed January 10, 2025, https://www.eia.gov/state/analysis.php?sid=WV.
7. U.S. Energy Information Administration, "New Mexico State Profile and Energy Estimates," *U.S. Energy Information Administration*, accessed January 10, 2025, https://www.eia.gov/state/analysis.php?sid=NM.

## 5. MEETING IN THE RADICAL MIDDLE

1. Miller, C.H.; Lane, L.T.; Deatrick, L.M.; Young, A.M.; Potts, K.A. "Psychological Reactance and Promotional Health Messages: The Effects of Controlling Language, Lexical Concreteness, and the Restoration of Freedom." *Human Communication Research*, Volume 33, Issue 2, 1 April 2007, https://doi.org/10.1111/j.1468-2958.2007.00297.x
2. U.S. Congress, Senate, *Congressional Report 116-114*, 116th Cong., accessed January 10, 2025, https://www.congress.gov/congressional-report/116th-congress/senate-report/114/1.

## 6. HOW TO BE CURIOUS ABOUT ENERGY STAKEHOLDERS

1. What does it take to modernize the U.S. electric grid? https://www.energy.gov/gdo/articles/what-does-it-take-modernize-us-electric-grid

## 7. HOW TO BE CURIOUS ABOUT THE ENVIRONMENTALISTS

1. California Air Resources Board, *Basics of Air Pollution*, accessed January 10, 2025, https://ww2.arb.ca.gov/sites/default/files/2020-09/basics-notes.pdf.

## 8. HOW TO BE CURIOUS ABOUT POLICYMAKERS

1. Nathan S. Balke and Mine K. Yücel, *Lifting the U.S. Crude Oil Export Ban: Prospects for Increasing Oil Market Efficiency*, Federal Reserve Bank of Kansas City, accessed January 10, 2025, https://www.kansascityfed.org/documents/468/Lifting_the_U.S._Crude_Oil_Export_Ban_Prospects_for_Increasing_Oil_Market_Efficiency06.pdf.
2. McKinsey & Company, "US Crude Exports Ban Lifted," *McKinsey & Company*, accessed January 10, 2025, https://www.mckinsey.com/industries/oil-and-gas/our-insights/us-crude-exports-ban-lifted.
3. Center for Strategic and International Studies, "The Economic Impact of Iran Sanctions," *CSIS*, accessed January 10, 2025, https://www.csis.org/analysis/economic-impact-iran-sanctions.
4. U.S. Environmental Protection Agency, "Inclusive Utility Investment Resource Library," *ENERGY STAR*, accessed January 10, 2025, https://www.energystar.gov/products/inclusive_utility_investment_resource_library.

## 10. AN "ALL OF THE ABOVE" SET OF ENERGY SOLUTIONS WILL MAKE THE WORLD A BETTER PLACE


1. Brian Evans, "High Electricity Prices Have Left 20 Million US Homes Behind on Their Utility Bill Payments," *Business Insider*, August 24, 2022, https://markets.businessinsider.com/news/commodities/high-electricity-prices-20-million-homes-behind-utility-bill-payments-2022-8.
2. Andrew Chatzky and James McBride, "'Made in China 2025': Is It a Threat to Global Trade?" *Council on Foreign Relations*, accessed January 10, 2025, https://www.cfr.org/backgrounder/made-china-2025-threat-global-trade.
3. David Song-Pehamberger, "Controlling Tomorrow: China's Dominance Over Future Strategic Supply Chains," *The Diplomat*, August 21, 2024, https://thediplomat.com/2024/08/controlling-tomorrow-chinas-dominance-over-future-strategic-supply-chains/.
4. Institute for Energy Research, "China Controls Global Rare Earth Mineral Trade," *Institute for Energy Research*, accessed January 10, 2025, https://www.instituteforenergyresearch.org/international-issues/china-controls-global-rare-earth-mineral-trade/.

# ABOUT THE AUTHOR

Dan Romito is a Managing Director overseeing the Consulting & Advocacy practice at Pickering Energy Partners. His career has centered on helping capital-intensive businesses efficiently navigate the ever-evolving capital markets and global regulatory landscape. He advises private companies, public issuers, and asset managers in the Energy, Industrials, Materials, Utilities, and Healthcare sectors to optimize the delicate balance between sustainability strategies and capital discipline. He pursues quality pools of capital, employs pragmatic sustainability-related directives, and executes efficient stakeholder management strategies.

www.ingramcontent.com/pod-product-compliance
Lightning Source LLC
LaVergne TN
LVHW050915210926
848482LV00012B/442
* 9 7 9 8 2 1 8 6 2 4 7 2 9 *